Electrostatic Ignitions of Fires and Explosions

Electrostatic Ignitions of Fires and Explosions

Thomas H. Pratt

BURGOYNE INCORPORATED

CONSULTING SCIENTISTS & ENGINEERS

MARIETTA, GEORGIA

CCPS CENTER FOR CHEMICAL PROCESS SAFETY

An AIChE Industry Technology Alliance

Center for Chemical Process Safety
of the
American Institute of Chemical Engineers
3 Park Avenue, New York, NY 10016-5991

Library of Congress Catalog Card Number: 97-093919

ISBN 978-0-8169-9948-4

CONTENTS

Figures

Tables

PREFACE

Albert Einstein once stated that a physical principle should be presented in its simplest form, but no simpler. In this regard, I have attempted to give the beginner only some simple electrostatic relationships which I deem important to the basic understanding of the subject. I have not followed his caveat and must plead guilty to the charge of oversimplification.

Most of the basic principles of electrostatics are quite simple but the rigorous mathematical treatment, which is dear to the physics professor's heart, can be very complicated. An effort has been made to give the beginner a few equations which can be used in a very broad brush approach in examining conditions for potential hazards in common industrial processes. For instance Gauss law which is the very bedrock of electrostatic theory is not even mentioned and only the direct fallout of Coulombs law (Equations 5, 6, & 7) is used to get the reader started in the understanding of electric fields. Other expressions are merely stated without a derivation from more basic principles. The notion here is to just give the reader the tool to do the work, but in so doing I perhaps run the risk that the tool may be misused on special occasions. Here, I would ask the reader to inquire further if there is any doubt in the use of the relationships contained herein.

Also, there are many cases where there are specific exceptions to some of the generalities. In this regard, I again am guilty of over simplification by not going into detail about many second or even third order effects. As an example, the text takes a global dielectric strength of air at 3×10^6 V/m even though it is well known that there are other considerations. Some texts discuss these considerations and get into Pachens law, homogeneous fields, breakdown with gap length, etc. I have opted to leave such discussions out of the running text unless they have some specific application to the examples and case histories.

In the running text an effort has been made to give references to definitive works where in depth discussions of a particular topic can be found. In many cases there are numerous places which could be referenced because some topics have been discussed many times by authors over the years, each with its own slant and bias. The ones I have cited are usually the ones where I learned of it or the ones I prefer when I need to look something up. Since authors refer to each other's previous works, there are many instances where data have been rattling around the literature for years and become conventional wisdom. For instance I have cited BS 5958 for Table 6.2 but it has been around since 1969 that I know of and perhaps sooner. Another conventional wisdom is that corona discharge is incendive to stoichiometric hydrogen/air mixtures. "Everyone" seems to agree, but a definitive reference to the experimental work, if there is one, has been lost in antiquity (Heidelberg, 1967 comes close)

The basic objective of writing the book was to educate the industry in the basics of electrostatics and to have a pseudo handbook of basic electrostatic data. Piecewise, there is very little original material in the text, figures, or tables; so if one has the need to obtain a copy of a particular item it is suggested that the original source(s) be consulted and cited. An exception to this are the three Nomograms of Chapter 9 which are original even though they were inspired by the nomogram of Bodurtha (1980) A blanket permission to copy Nomograms 9.1, 9.2, & 9.3 in any form is granted to whomever may need copies of them for any purpose. It is requested however that Burgoyne Incorporated, Marietta, Georgia be cited as having given permission.

In describing the evolution of an electrostatic charge from its genesis to an ignition, the terms *generation, accumulation,* and *discharge* are used in some texts and standards while in others the terms *separation, accumulation,* and *discharge* are preferred. There is no unanimity of agreement between authors for using the term *generation* or *separation*, and sometimes heated and adamant discussions ensue. It is left to the serious student of electrostatics to sort out his own preference and enjoy joining the fray.

A special word of thanks is given to my mentor Dr. George M. Williams for reviewing the text, not that he fully condones my broad brush approach to approximating electrostatic problems, but that he has kept me honest by not letting me get too far afield and keeping me straight in the use of the term *separation* throughout the text.

Chapter 1

Basic Concepts

Nutshells:

[1] Removal of an electron from a molecule leaves a positively charged ion - a unit positive charge.

[2] Attachment of an electron to a molecule creates a negatively charged ion - a unit negative charge.

[3] Like charges repel each other; unlike charges attract each other.

[4] Charges move about freely on the surfaces of conductors.

[5] Insulative materials resist the movement of charges - either across their surfaces or through their interiors.

[6] An electric field is a region in space where electric forces can be experienced.

[7] A dielectric is a insulating material which will permit the passage of an electric field.

[8] Electrostatic discharge or breakdown occurs when the electric field between two electrodes exceeds the critical value of the dielectric breakdown strength of the intervening material.

[9] All materials have some conductivity. Errant charges will therefore dissipate or recombine if given enough time.

[10] In order to have an electrostatic scenario for the ignition of a fire or an explosion, four conditions must be satisfied: Separation, Accumulation, Discharge, and Ignitable Mixture.

1.1 The Electrostatic Charge

An electrostatic charge is a result of a large quantity of ions of the same polarity being accumulated in the same region at the same time. An accumulated electrostatic charge will result in an electric field.

1.1.1 Electrons, Protons, and Ions

Atoms are the building blocks of all matter. An atom can be viewed as a positive nucleus surrounded by a miniature solar system of negative electrons. In its normal state the number of negative charges (electrons) in the orbits around the nucleus equals the number of positive charges (protons) in the nucleus so that the atom, or the molecule in which it is contained, is in an uncharged or electrically neutral state. Molecules are made up of atoms and if an electron is removed from a molecule, that molecule will carry a positive charge. If the electron which was removed from one molecule attaches itself to another molecule, the second molecule will carry a negative charge. It follows then, that in a closed system, charge can only be separated - not created. For every positive charge in a closed system there must also be the equal and opposite negative charge somewhere else in the system. Charge is measured in coulombs, one coulomb contains 6.24×10^{18} unit charges (electrons).

Charges of opposite polarities attract each other and charges of like polarities repel each other; therefore, as charge separation is occurring, there are forces which work toward charge neutralization since the positive nuclei will be attracting the errant electrons. Charge separation occurs when the forces of nature exceed the attractive forces between an electrons and its positive nuclei. If the charges are free to move through a conductive medium, then the attractive forces will prevail and charge will not be separated. On the other hand if charges are caught in an insulative medium they are inhibited in their movement and charge separation can occur. If all matter were made up of perfect conductors of electric charges, then the attractive forces would prevail and charge separation would not occur. In most cases it will depend upon how "conductive" the medium is whether charge separation occurs or not. It can be seen then that the concepts of resistance, resistivity, conductance, and conductivity are important in the understanding of static electricity.

In the discussions of electrostatics the adjectives "conductive", "dissipative", and "insulative" (and their equivalent nouns) are used in a semi-quantitative sense to allude to the electrostatic properties of materials. Some investigators use the terms "nonconductive" and "nonconductor" to refer to those materials having very high resistivities. Since all materials have some conductivity, it is a bit of intellectual dishonesty to use these terms. Perhaps the term "poor conductor" is more to the point, but the terms "insulative" and "insulator" will be used herein, Table 1.1.

Table 1.1: Nomenclature for Resistivity

	Volume Resistivity, ρ $\Omega \cdot m$	Surface Resistivity, λ Ω/square
Conductive	$\rho < 10^2$	$\lambda < 10^5$
Dissipative	$10^2 \leq \rho < 10^9$	$10^5 \leq \lambda < 10^{12}$
Insulative	$\rho \geq 10^9$	$\lambda \geq 10^{12}$

Electrostatic Discharge Control Handbook, 1994
Electrostatic Shielding Materials have volume resistivities less than 1 $\Omega \cdot m$.

Any conductor (super conductors excepted) will resist the flow of charges through it. This resistance can be measured by pushing a current of electricity through the conductor in question. Ohm's law states that the amount of current flowing through a conductor will be directly proportional to the potential (voltage) across the conductor and inversely proportional to the resistance of that conductor.

$$I = \frac{V}{R} \tag{1}$$

I ≡ Current, A
V ≡ Potential, V
R ≡ Resistance, Ω

Resistance is not an intrinsic property of a material since the resistance of a conductor will be a function of its dimensions. Resistivity is an intrinsic property of a material and is defined in terms of the

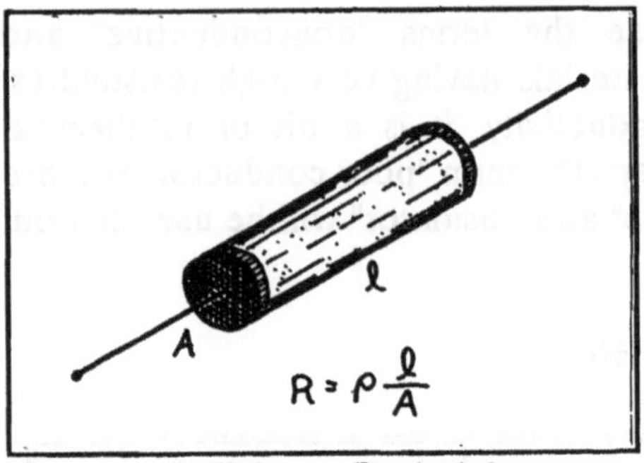

Figure 1.1: Volume Resistivity

resistance, length, and cross section of a body of the subject material. The resistance of a conductor of constant cross section is directly proportional to its length and inversely proportional to its cross sectional area, Figure 1.1. The proportionality constant is the resistivity of the material from which the conductor was made and is expressed in units of ohm meters.

$$R = \rho \frac{\ell}{A} \qquad (2)$$

$\rho \equiv$ Resistivity of material, Ω·m
R $\equiv$ Resistance of a conductor made from the material, Ω
$\ell \equiv$ Length of the conductor, m
A $\equiv$ Cross sectional area of the conductor, m^2

Conductivity is the reciprocal of resistivity and is expressed in units of siemens per meter.

$$\kappa = \frac{1}{\rho} \qquad (3)$$

$\kappa \equiv$ Conductivity, S/m

Notice that the conductivity and resistivity between metals and plastics can be 27 orders of magnitude, Table 1.2. It is therefore to be expected that the qualitative and quantitative characteristics of static electricity in plastics are much different than those of current electricity in metals.

Table 1.2: Comparison of Electrical Properties of Metals and Plastics

Material	κ, S/m	ρ, Ω·m
Metals; eg., Copper	10^{8}	10^{-8}
Plastics; eg., PTFE	10^{-19}	10^{19}

See also Table 9.1

In real life situations, when charge separation occurs the charges will accumulate either on the surface of a material or within the bulk of a material. If an amber rod is rubbed with silk, a charge will be "stuck" to the surface of the rod and the dissipation of that charge will be governed by the *surface conductivity* of the amber. If an insulative liquid is vigorously pumped into a tank, a charge will reside within the bulk of the liquid and the dissipation of that charge will be governed by the *bulk conductivity* of the liquid.

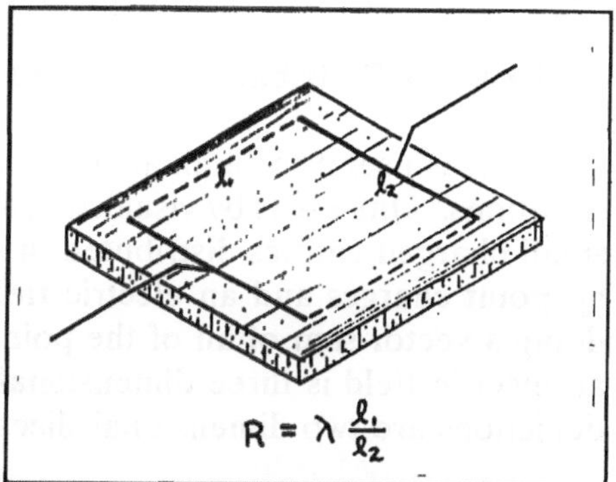

Figure 1.2: Surface Resistivity

In cases where the bulk conductivity is very low, currents may flow across the surface of the material rather than through the body of the material providedthe surface is sufficiently conductive, i.e., essentially all of the current flows across the surface and none through its interior. One can then visualize the current as flowing across a rectangle of the surface of an insulator which has electrodes on its opposite edges. The resistance to the flow will be directly proportional to the length of the rectangle, ℓ_1, and inversely proportional to its width, ℓ_2, and the constant of proportionality is the surface resistivity, λ, Figure 1.2. It can be seen then that a square of a material will have a constant resistance no matter the size of the square, and since the unit for surface conductivity is the ohm, surface conductivity is sometimes quoted in terms of the units of "ohms per square".

$$R = \lambda \frac{\ell_1}{\ell_2} \tag{4}$$

$R \equiv$ Resistance, Ω
$\lambda \equiv$ Surface resistivity, Ω
$\ell_1 \equiv$ Length, m
$\ell_2 \equiv$ Width, m

When firs presented with the units of ohms per square, many students either think it is a typographical error or ask the question "Ohms per square what?" But after a moments reflection about how surface conductivity is determined, then the units make sense. A square of material is selected, if a small square is used, the electrodes are short and close together; if a large square is used, the electrodes are longer but further apart. In both cases the resistance is the same, thus the units of "ohms per square". Obviously, a determination of λ is not made with a configuration of electrodes as shown in Figure 1.2 since there would be all sorts of edge effects. The determination is made with electrodes which are concentric circles and λ is derived from the geometry.

1.1.2 Charge Distribution; Point, Space, and Surface Charges

It is useful to think about how charges get distributed in practical situations. One starts by visualizing a charge accumulation as a set of very small charged entities distributed in space. These are then considered to be point charges and an electric field can be constructed from them by doing a vector sum of all of the point charges. One must remember that an electric field is three dimensional and that two dimensional graphical depictions are two dimensional slices of a three dimensional field.

A *space charge* exists when a charged, insulative material occupies a space. The electrostatic charge is bound to the insulator and the insulator is physically confined to the space in question. Examples are a charged powder in a silo and a charged liquid in a tank. Another example is a charged mist but the droplet can be a conductor, it is the intervening material (air) that is the insulator. In these examples there is an electric field within the space charge.

A *surface charge* exists when an electrostatic charge resides on the surface of a insulative material. The charge accumulates on the surface of the material rather than in its interior. An example of a surface charge is a sheet of plastic which has been rubbed by a cloth. The accumulated charge is on the surface of the plastic sheet.

1.2 The Electric Field

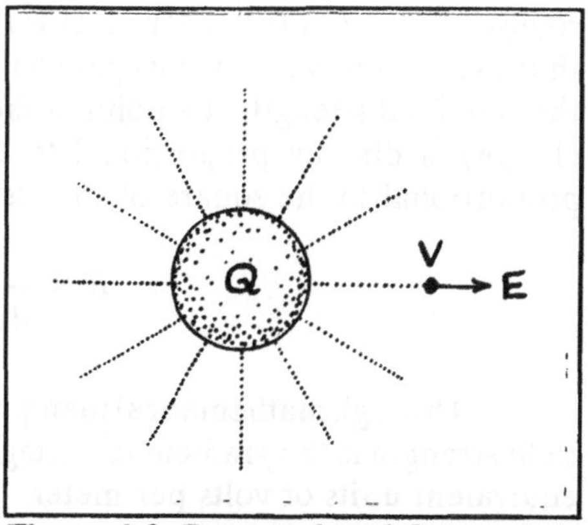

Figure 1.3: Potential and Strength of an Electric Field

A good way to visualize what is going on in a particular electrostatic situation is to visualize the electric field created by the accumulated charge. A way to begin is to visualize some ideal cases and then extrapolate those cases to real life situations. As in all of the physics books, this is done by first considering a charged spherical conductor in free space, Figure 1.3. There will be a spherically symmetric electric field about the sphere. An *electric field* is a region in space where electrostatic forces can be experienced. If a very small test charge (like one electron) is placed in the field, there will be a force on the test charge and it will take some energy (force times distance) to move it about.

That is, energy is required to bring the test charge from infinity to a point in the electric field. Then, the potential energy on the test charge in the electric field is used to define electric potential. The point where the test charge resides in the electric field has a *potential,* V, which is defined as the energy per charge on the test charge that was required to move it to that point. The potential, V, therefore has the units of energy per charge on the test charge; however, since potential is a very basic parameter it has been assigned the unit of the volt. From Coulombs Law it follows that the voltage at a point in the electric field of a sphere (or a point charge) is directly proportional to the amount of charge and inversely proportional to the distance (radius) from its center.

$$V = \frac{Q}{4\pi e e_0 r} \quad (5)$$

As the test charge resides at the point in the electric field there will be a force on the test charge. The magnitude of the force on the test charge is the measure of the strength of the electric field at that point and is designated by E. Note that E is a vector quantity since it has both magnitude (the strength of the field) and direction (the direction of the force on the test charge), and that is enough said about vectors. The *field strength*, E, is defined in terms of force per charge on the test charge and thus has the units of newtons per coulomb. From Coulombs Law it follows that the field strength at a point in the electric field of a sphere (or a point charge) is directly proportional to the amount of charge and inversely proportional to the square of the distance (radius) from its center.

$$E = \frac{Q}{4\pi\epsilon\epsilon_0 r^2} \tag{6}$$

Through mathematical manipulation and the laws of physics, electric field strength is the gradient of voltage in the electric field and thus has the equivalent units of volts per meter.

$$E = -\frac{\partial V}{\partial r} \tag{7}$$

E ≡ Electric field strength, V/m
V ≡ Potential, V
Q ≡ Charge, C
$\epsilon\epsilon_0$ ≡ Permittivity, F/m
r ≡ Radius, m

The charged sphere of Figure 1.3 can be considered in several ways. First consider it as a charged solid hunk of metal. All points on the surface of the sphere are at the same potential and thus form an *equipotential surface*. As a consequence of Gauss's Law all of the charge on the sphere are on the outer surface. The profile of the potential and field strength of the electric field about the sphere is depicted in Figure 1.4. Notice that the potential inside the sphere is everywhere the same. This leads

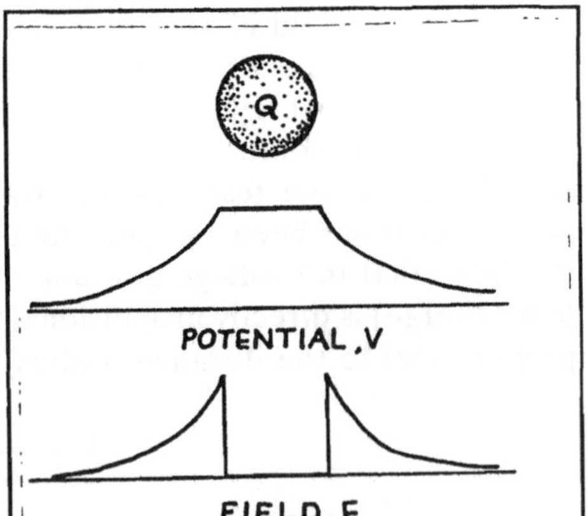

Figure 1.4: Profile of Field Strength and Potential about a Sphere

to the very important fact that there is no electric field inside of the sphere where the gradient is zero.

Now consider a hollow sphere which has the same charge. Again, all of the charge is on the outer surface and there is no field within the void space of the center. Therefore, a charge can be brought into the interior of the sphere and deposited on the inside surface where it will immediately move to the outer surface. This is the principle of the Van de Graaff generator where charge is carried by a moving belt to the interior of the electrode, Figure 1.5. There is no field on the interior of the metal electrode to resist the further addition of charge; therefore, a continuous stream of charge can be brought into the interior of the metal enclosure where it will accumulate on the outer surface. Tremendous amounts of charge can be accumulated in this manner and is important in a number of process situations which will be developed later.

1.2.1 Mapping Electric Fields

The visualization of an electric field can then be done in terms of the field strengths and the potentials at various points in the field. Graphical representations can be constructed by drawing equipotential surfaces and field lines.

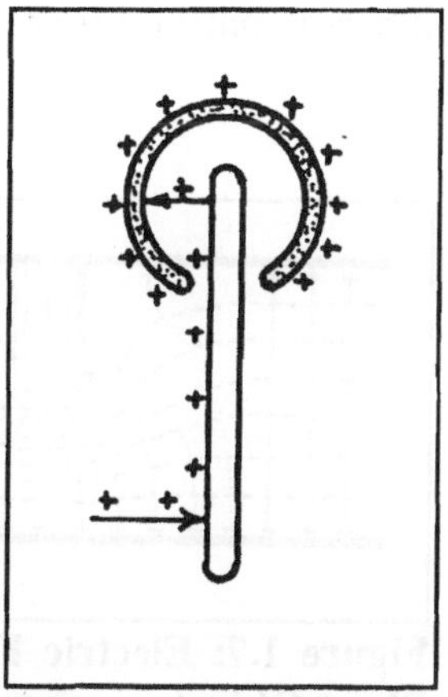

Figure 1.5: Van de Graaff Generator

It is obvious that each point in space will have a unique potential; therefore, all points having the same potential will define an equipotential surface; i.e., a surface which is everywhere at the same potential. Likewise, it can be seen that the force vector at any point on an equipotential surface is perpendicular to the surface. An imaginary field line can then be constructed through each of the successive equipotential surfaces while maintaining perpendicularity. This will define a field line through the electric field. A field line can also be visualized as the trajectory a single electron would take if it were, in imagination, inserted into the field and let loose. A field line can also be thought of as beginning on an entity of positive charge and ending on an equal and opposite entity of negative charge.

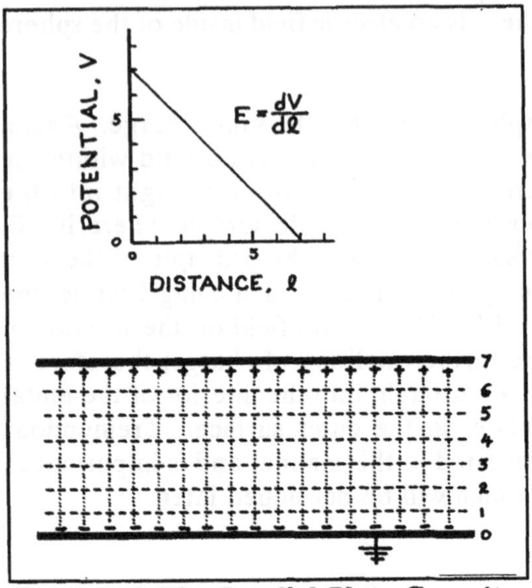

Figure 1.6: The Parallel Plate Capacitor and Homogeneous Electric Field

A basic geometry to be visualized is that of a parallel plate capacitor; i.e. two parallel planes each carrying an equal and opposite electrostatic charge, Figure 1.6. Between the plates there will be an electric field; i.e., a region in space where electric forces can be experienced. The equipotential surfaces are parallel planes with perpendicular field lines (edge effects are neglected). Note that the field strength, E, is everywhere the same since the slope or gradient of the potential, V, between the plates is constant. This is the case of the homogeneous electric field. Note that an increase in voltage between the plates has with it a concomitant increase in field strength.

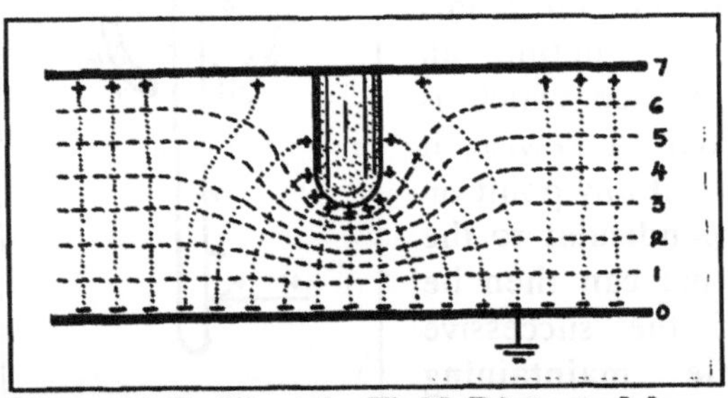

Figure 1.7: Electric Field Distorted by a Blunt Object

When objects are inserted into an electric field, the field is distorted. Consider the addition of a rounded conductive object to one of the plates of the parallel plate example, Figure 1.7. The equipotential surfaces are squeezed closer together making the slope or gradient higher and increasing the field strength between the object and the other plate. Also note that the field lines are "collected" by the inserted object.

Consider the addition of an object with a sharp point rather than a rounded one, Figure 1.8. Again the field is distorted and the most distortion occurs at the tip of the pointed object. It is important to note that the field lines are very close together at the tip of the object and thus the field strength is very high at this point. When all else is equal, the field strength at a projection or sharp corner of a conductive object is inversely proportional to the radius of the object. Thus, for small radii at tips or sharp corners there is a high field strength. The significance of this observation will be revisited in the discussions of corona discharge.

Consider a conductive object levitated in an electric field, Figure 1.9. The object distorts the electric field and charge separation occurs on the surface of the conductive object. There is conservation of charge in that there is just as much positive charge as there is negative charge on the object. Being an equipotential surface, the object is at the potential of the electric field even though there is no net charge on the surface. The object is polarized in that charge separation has occurred and the process is termed *induction*, cf. ¶ 2.11.

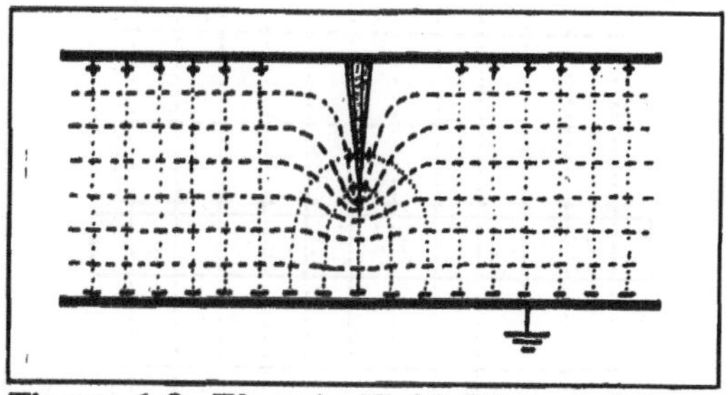

Figure 1.8: Electric Field Distorted by a Sharp Object

Where there is a charged insulative surface, there will be an electric field both above and below the surface. The field can be measured at either surface; and, theoretically, does not decreases as the distance from the surface increases, Figure 1.10. However, in practice a field meter will indicate a reduction in the field strength as it is moved away from the surface because the field meter distorts the field; cf., Chapter 8.

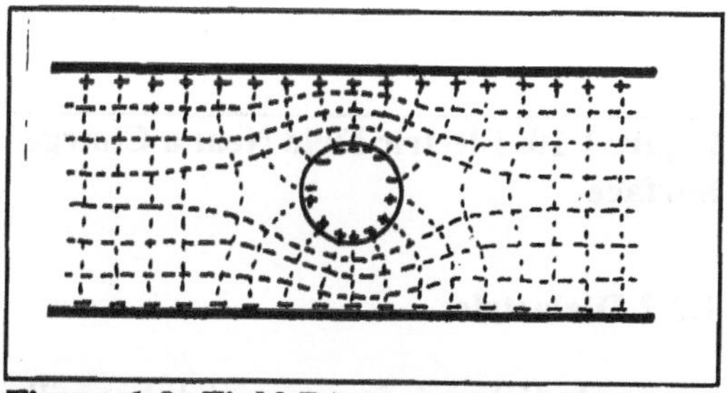

Figure 1.9: Field Distorted by a Levitated Sphere

Consider a metal tank filled with a charged mist such as would be the case after it was washed or steamed, i.e., a space charge. There will be an electric field within the tank, Figure 1.11. The shell of the tank will be an equipotential surface at ground potential (it is sitting on the ground)

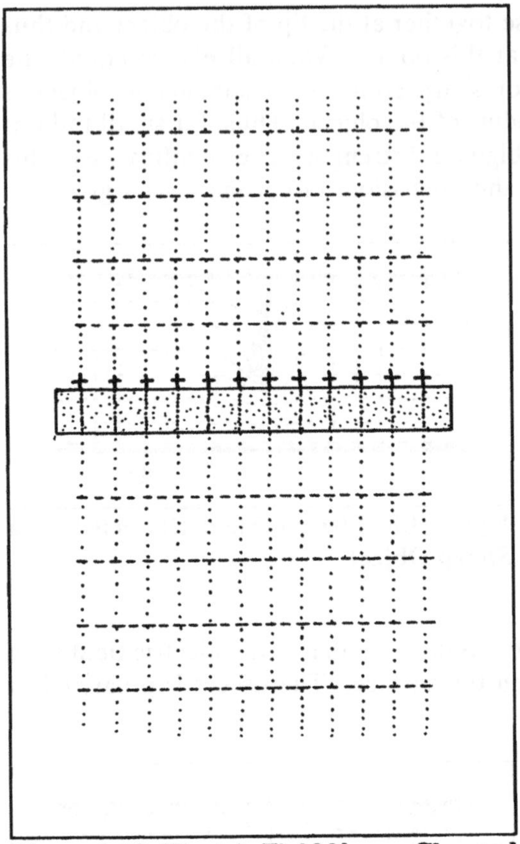

Figure 1.10: Electric Field Near a Charged Surface

and there will be equipotential surfaces of higher potential within the mist. The potential will increase from zero at the shell to its maximum near the center, Figure 1.12. Remembering that the field strength is the gradient of potential, Equation 7, it can be seen that the field strength will be a maximum at the shell and zero at the center as indicated by the slopes in Figure 1.12. Assuming the mist to be uniformly charged and all of the droplets being mutually repulsive, the field strength at any point can be visualized in terms of the force on a droplet of mist at that point. There is no net force on a droplet at the center but there is a maximum force on a droplet at the shell and an intermediate force at points in between.

1.2.2 Dielectrics

A dielectric is a material that will pass an electric field. In the case of the parallel plate capacitor, Figure 1.6, where the plates are separated by an insulating material, the capacitance will depend upon the dielectric properties of the intervening material. The capacitance will depend upon how the field is "permitted" to exist by the material which is between the plates. This is termed the *permittivity* of the material, and the baseline permittivity to which all other materials are referred is that of a vacuum, ϵ_0. When expressed in SI units the permittivity of a vacuum, ϵ_0, becomes

$$\epsilon_0 = 8.85 \times 10^{-12} \text{ F/m}$$

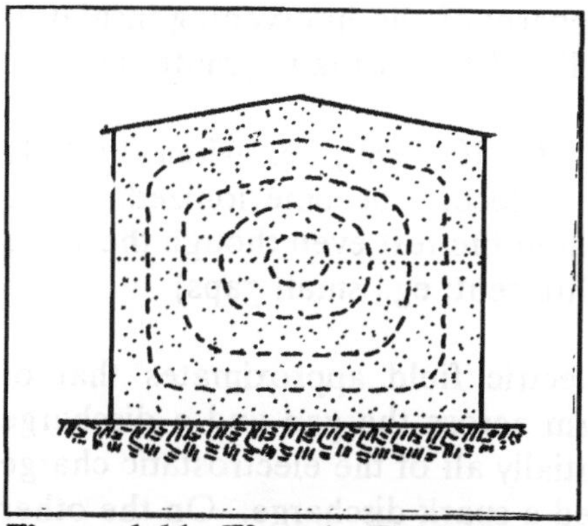

Figure 1.11: Electric Field from a Space Charge within a Tank

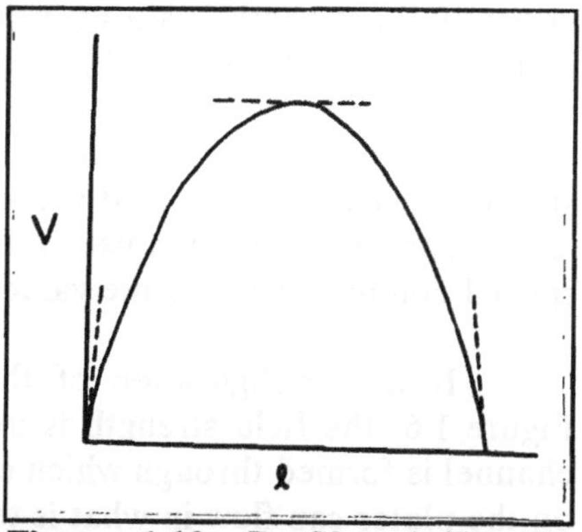

Figure 1.12: Profile of Electric Field Through Center of Tank

The units of permittivity are usually expressed as farads per meter but in other mathematical relationships it has the equivalent units of seconds per ohm meter, coulombs per volt meter, etc.

The dielectric constant, ϵ, of a material is defined as the ratio of the permittivity of that material to that of a vacuum. Thus the permittivity of a material is written as $\epsilon\epsilon_0$ where ϵ is the dielectric constant and ϵ_0 is the permittivity of a vacuum.

The dielectric constant for air is one since air "permits" the passage of an electric field with almost no interference, but the dielectric constant of a metal is infinity since it completely attenuates the passage of an electric field. The dielectric constant of a material is always greater than unity and can be taken as a measure of how much a particular material will attenuate an electric field. Also, there is a rough correlation between dielectric constant and conductivity for a given material.

1.2.3 Dielectric Breakdown

When an electric field exists in an insulating medium, the electrons in the molecules of the medium will be acted upon by the electric field. When the field strength is increased, a point is reached where electrons are ejected from their orbits and a breakdown of the insulating properties of the medium takes place. When this occurs a current flows through the plasma until the electric field collapses and the current ceases to flow. Breakdown occurs when the field strength reaches a critical value. This critical field strength is an intrinsic property of the intervening insulator, and it is termed the *dielectric strength*, $\mathbf{E_b}$, of the insulating material.

The dielectric strength for air is 3×10^6 V/m; i.e., when the field strength reaches this critical value air molecules become ionized. (This value is an approximation used throughout the text even though there are special conditions where the value is different; eg., small gaps)

If the configuration of the electric field approximates that of Figure 1.6, the field strength is uniform across the gap and a discharge channel is formed through which essentially all of the electrostatic charge on the plates can flow in what is termed a spark discharge. On the other hand, if the intervening medium is air and the electric field approximates that of Figure 1.8, the field strength exceeds the dielectric strength of air only at the tip of the electrode. In this case ions are formed only in the region of the tip and a discharge channel is not formed; however, the ions migrate through the air to their opposite counterpart in what is termed corona discharge.

When a combustible material is mixed with air in the gas phase a combustion reaction can take place providing the concentrations are within the flammable limits and there is an ignition source. An optimum fuel/air mixture can remain in a quiescent state forever as long as there is no ignition source. But, when ignition takes place a fire or explosion results.

Take for example the simple reaction of methane with atmospheric oxygen:

$$CH_4 + 2\ O_2 \rightarrow CO_2 + 2\ H_2O$$

The mechanism for the reaction is not that one molecule of methane wads up with two molecules of oxygen to come apart as one molecule of carbon dioxide and two molecules of water. There are a whole series of complicated intermediate chemical reactions involving a variety of molecular fragments occurring in a chain branching mechanism during the combustion reaction. In order to get things started, a critical concentration of molecular fragments must be exceeded, both in space and in time. One way to do this is by electrostatic discharge, but the spark must be intense enough and long enough to create the necessary molecular fragments to get things started.

This leads to the notion of a minimum ignition energy, MIE, for ignition by spark discharge. That is, there is a critical spark energy above which ignition will be effected for a given fuel/air mixture and below which ignition will not take place. In the extreme, this view is over simplified but it suffices for most practical purposes.

Some gasses and vapors can also be ignited by brush discharge and cone discharge but in these cases discharge energy cannot be reckoned and ignition criteria are difficult to establish. In these cases, MIEs can only be used as crude approximations, at best.

The mechanisms for ignition of dust (and mist) suspensions is much more complicated than that for gasses and vapors. In general, it takes more energy to get the combustion reaction started.

1.3 Ground Potential

An item is said to be at ground potential when its potential is that of the ground or the potential of the earth. As it was shown in Figure 1.7, the lower plate with the negative charge was at ground potential even though it held the negative charge entities. The negative charge entities are present in the grounded plate because of the attraction by the positive charge entities in the upper plate. Thus, the field lines begin at a positive charge entity and end at an equal and opposite negative charge entity.

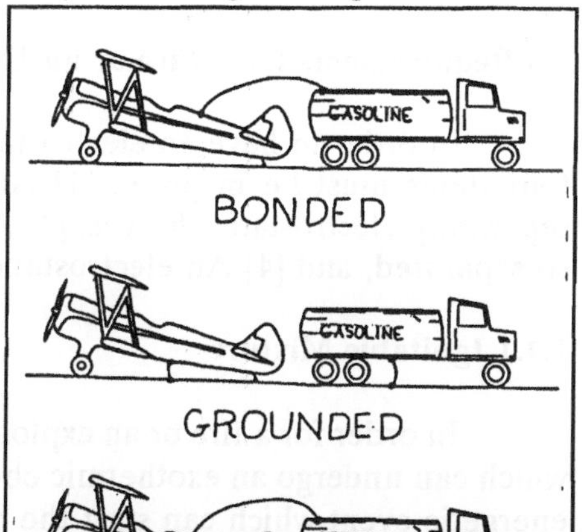

Figure 1.13: Bonding and Grounding

1.3.1 Grounding

The planet earth can be considered an infinite source or sink for electrostatic charges; they may flow from the earth or to the earth when they have a conductive path to do so. When such a conductive path is established between a conductive object and the earth, either by design or by happenstance, the object is said to be *grounded*. (In this regard the British nomenclature of saying that the object is *earthed* may perhaps be preferred.) In any case, *ground potential* is zero potential or voltage and any conductor electrically connected to it will likewise be at zero potential or voltage. Sparking cannot occur between two grounded, conductive objects.

1.3.2 Bonding

If two conductive objects are electrically connected together, they are said to be *bonded*. For example, during the refuelling of an aircraft where a wire is connected between the fuelling truck and the aircraft, they are said to be *bonded* if there is no conductor provided to *ground*. (Both the truck and aircraft are on rubber tires and may be on a insulative surface, so they may not be *grounded*.) If there is an electrical connection between the aircraft and ground *and* if there is an electrical connection between the truck and ground, the units are said to be *grounded*, Figure 1.13. In the case of bonding, the units are maintained at the same potential and there can be no spark between them. *BUT*, they may not be at ground potential and there could be scenarios where there could be a spark between the units and ground! In such operations redundant grounding and bonding is usually recommended by operating companies.

1.4 Requirements for a Fire or an Explosion

In order to have an electrostatic ignition of a fire or an explosion, four items must be in place: [1] An ignitable mixture, [2] A means of separating electrostatic charges, [3] A means of accumulating the charges so separated, and [4] An electrostatic discharge in the ignitable mixture.

1.4.1 Ignitable Mixture

In order for a fire or an explosion to occur, there must be a mixture which can undergo an exothermic chemical reaction and there must be an energetic event which can start the chemical reaction in the mixture. In the general case some sort of fuel is mixed with some sort of oxidizer and a reaction is started in the resulting mixture by some sort of an ignition source; i.e., fuel, oxidizer, and ignition. It should be noted that the fuel and the oxidizer can stay mixed for an infinite period of time without reacting. It is when there is an ignition source to start the reaction that the fire or explosion occurs.

We usually think of a fire as a material burning in air. This is the most common type of fire and it is where atmospheric oxygen reacts with some sort of fuel to give a flame. When the fire is very rapid we think of it as an explosion. When a combustible substance, such as a dust, is dispersed in air, conditions can be such that a combustion reaction can occur if ignited; many times these are explosive mixtures. On the other hand when a solid fuel and a solid oxidizer are mixed together, we have an explosive. There are of course other combinations but the present discussions primarily deal with fuel-air mixtures and explosives.

1.4.2 Separation

Electrostatic potentials can be separated by at least five mechanisms: (1) contact and frictional charging, such as rubbing a silk cloth over a glass rod, (2) double layer charging, such as a liquid flowing through a pipe, (3) induction, such as a charged surface inducing another charge on an adjacent ungrounded conductor, (4) charge transfer, such as when a charged object contacts an uncharged object and the charge is then shared between them, and (5) corona charging, such as impressing a charge on the drum of a copy machine.

1.4.3 Accumulation

As soon as a charge is separated (i.e., unit charges separated one from another), the forces of mutual repulsion between the like ions which make up the charge act to dissipate the charge. If the charge is in a conductive medium, the charge will easily dissipate through the medium to ground; but if the medium is insulative, the charge cannot easily dissipate by finding its way to ground. In this manner charges are accumulated somewhere in a process and there is an electric field associated with the accumulated charge.

1.4.4 Discharge

The accumulation of electrostatic charge has with it the accumulation of energy somewhere in the system. When the charges are accumulated, the system can accommodate to the accumulation only so far. When the ability of the system to accommodate to the increase of charge is exceeded, the accompanying electric field will cause ionization and subsequent breakdown of the accumulation. This breakdown usually occurs in spurts or electrostatic discharges and may be classified into six types: [1] corona discharge, [2] brush discharge, [3] spark discharge, [4] bulking brush discharge, [5] propagating brush discharge, and [6] lightning.

Chapter 2

Separation and Accumulation of Charge

Nutshells:

[1] When the rate of charge separation exceeds the rate of charge dissipation, potentials may increase until discharges occur.

[2] Charge can usually be considered to have dissipated after five time constants. (Haase, 1977)

2.1 Mechanisms of Charge Separation

Electrostatic potentials can be separated by at least five mechanisms: [1] contact and frictional charging, such as rubbing a silk cloth over a glass rod, [2] double layer charging, such as a liquid flowing through a pipe, [3] induction, such as a charged surface inducing another charge on an adjacent ungrounded conductor, [4] charge transfer, such as when a charged object contacts an uncharged object and the charge is then shared between them, and [5] corona charging, such as impressing a charge on the drum of a copying machine.

The mechanisms of charge separation and charge accumulation are so interwoven it is sometimes difficult to keep them logically separated. As the above mechanisms show, charge separation and accumulation go hand in hand. One way to look at it is that separation is on a molecular scale where opposite charges become aligned in some fashion, and accumulation is on a macroscopic scale where some external force pulls a number of the aligned charges apart and puts them in one place. In frictional charging the charges aline themselves on a molecular scale and the movement of the materials pull them apart. The same can be said for double layer charging. As will be developed later, induction is a composite mechanism. Charge transfer is a straightforward mechanism and there is no separate discussion except where it is included as an example. Corona charging is usually intentional and is not developed.

2.2 Charge Alignment

Whenever two dissimilar materials come into contact the forces of nature cause the molecules to align themselves so that either the positive or the negative portion of the molecule orients itself toward the interface. The interface may be between two solids, between a solid and a liquid, or between two immiscible liquids. Charge orientation also occurs in liquids at a liquid/gas interface, but there is no charge orientation in the gas.

There are two cases which are of particular interest where charge alignment is important in the formation of electrostatic potentials - *contact and frictional charging* and *double layer charging.*

2.3 Contact and Frictional Charging

Contact and frictional charging has been known for centuries but it is perhaps the least understood of the electrostatic phenomena today. This phenomena is known as triboelectricity - the creation of an electric charge by friction. Frictional charging takes place at solid-solid interfaces. For many solids it is only necessary to touch them together and pull them apart to get some charge separation; however, if the surfaces are rubbed together, a higher surface charge density will result. An oversimplified way of looking at frictional charging is that the electrons are rubbed off of one surface and attach themselves to the other. Thus, one surface will carry a positive charge while the other will carry a negative one. Most frictionally charged surfaces have areas of both positive and negative charges; the net charge of the surface is determined by the one which predominates. But, when the materials are electrically very different, there will be very little, if any, areas of opposite polarity.

2.3.1 Surface Charging

Surface charging is only evident in poor conductors because the charges are not free to move and will remain on the surface of the material. They can therefore be separated from one another as the materials are moved apart since they remain on the surface. If similar attempts to separate charges are made with two good conductors, such as metals, the electrons are free to move through the metal and will not remain on the surface; therefore, they will not be separated from one another when the materials are rubbed or pulled apart. However, surface charging can result between a conductive surface and the surface of a poor conductor.

When rubbing is vigorous, there can be significant charge accumulation, even to the point where incendive discharge can occur (Gibson and Lloyd, 1965). In some cases the density of surface charges can become quite high; but since the surfaces are in intimate contact, the electric field is primarily between the layers. In this manner surface charge densities can exceed those of a single surface in air. When such highly charged surfaces are rapidly separated, discharges will occur.

2.3.2 Powder Charging

Any time powders are moved about in a materials handling process there will be charge alignment at the interfaces; both between particles and between particles and process equipment. Many times the positive and negative charges remain next to each other so that there is little if any external electric field. But, when the material handling operation moves the powder about, charges are separated - sometimes in tremendous quantities. Charge separation is then to be expected in powder handling operations such as sieving, pouring, scroll feeding, grinding, micronizing, chuting, conveying, etc.

2.4 Double Layer Charging

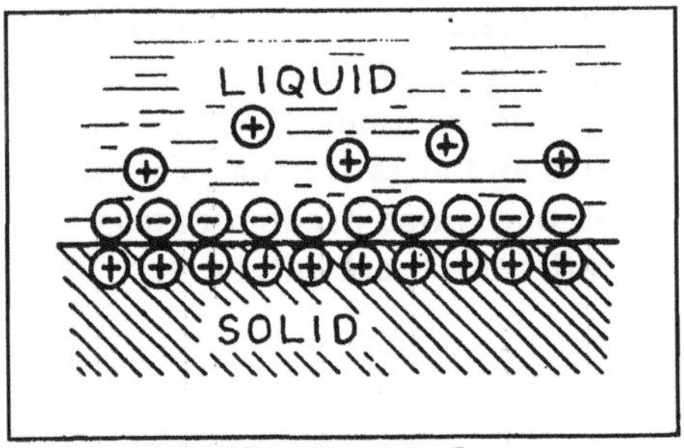

Figure 2.1: The Double Layer

Double layer charging results from charge separation which occurs on a microscopic scale at liquid interfaces; solid-liquid, gas-liquid, or liquid-liquid. Molecules within the liquid tend to aline themselves at the interfaces, with one charge being held at the interface while the portion of the molecule carrying the opposite charge extends into the liquid. Because of this alinement, an electric field is created at the interface and charges are induced onto molecules in the adjacent layer of liquid - the double layer. There have been several models suggested for the ionic configuration of the double layer (Cross, 1988), one of which is shown in Figure 2.1. If the liquid is moved, charge separation occurs where one charge will move along with the moving liquid and the opposite charge will be left at the interface. This is the mechanism of *streaming currents* in pipes which carry insulative liquids.

2.5 Charging of Drops, Mists, and Aerosols

Blanchard [1963] has shown that the breaking of air bubbles at an air-water interface result in the production of highly electrified water droplets. At gas-liquid interfaces, there is a double layer of charged molecules at the surface of the liquid. As the bubble comes to the surface, a thin layer of liquid is formed over the bubble. As the bubble rises to the surface and bursts the dimensions of the droplets which are formed is less than that of the double layer resulting in the "pinching off" of a bit of charge in each droplet, Figures 2.2 and 2.3. The forces of surface tension are greater than the repulsive electrostatic forces as the droplets are formed. Depending on conditions, either a positive or negative space charge can be formed above the surface of the liquid. [In the case of sea water the charge on the mist is positive (Blanchard, 1963)].

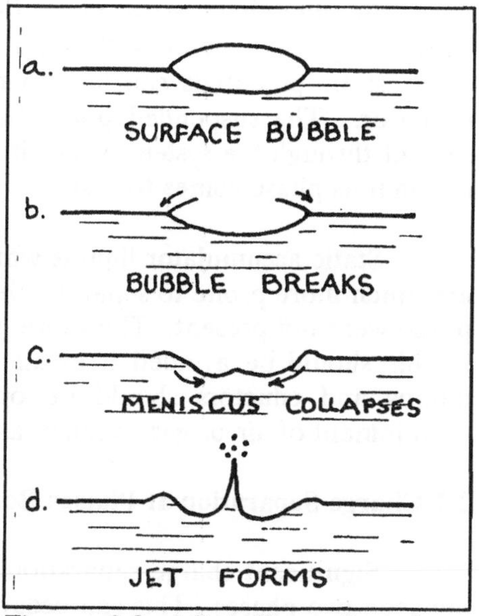

Figure 2.2: Mist Formation from Bursting Bubbles

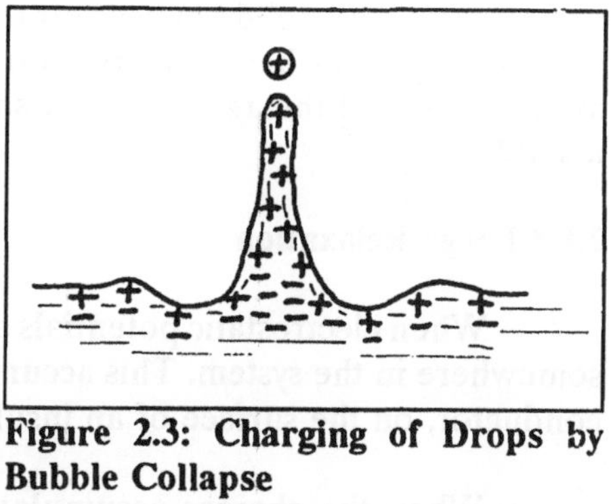

Figure 2.3: Charging of Drops by Bubble Collapse

One would expect that the space charge above the liquid would impede the additional formation of charged droplets at the surface of the liquid so that only minimal electric fields would be formed. On the contrary, it has been found (Nifuku, Vonnegut, and Blanchard, 1977) that fields of the order of 100 Kv/m produce little effect on the electric charge carried by the ejected liquid droplets. So from the viewpoint of charging, significant space charges and gradients can be formed at the surface of the liquid under favorable conditions.

2.6 Two Phase Flow

Anytime there is two-phase flow (i.e., a gas with solid particles or liquid drops or a liquid with solid particles or gas bubbles) very high electrostatic potentials can be separated if the continuous phase is an insulator. The suspended phase becomes charged as it moves and is carried through the system where it can become accumulated where the continuous phase comes to rest.

Static accumulator liquids which contain an entrained gas or solid are much more prone to separate electrostatic charges than if the second phase were not present. Therefore, two-phase flow of static accumulator liquids should be avoided if at all possible. As an example, stripping pumps and eductors should be operated in a manner to avoid the entrainment of air or gas as much as possible.

2.7 Charge Separation at Phases Boundaries

Significant charge separation can occur when there is movement between two phases. This can occur when there are gas/liquid, gas/solid, liquid/liquid, or liquid/solid interfaces. Examples of these in industrial processes are, respectively: [1] blowthrough or the "soda straw" effect where droplets of liquids are formed in a stream of gas, [2] pneumatic transport where a solid bounces off the walls of the conveying duct, [3] two immiscible liquids being pumped through a pipe or one settling out from another, and [4] the agitation of a slurry in a mixer or the settling out of a solid.

2.8 Charge Relaxation

When electrostatic potentials are separated, the charges accumulate somewhere in the system. This accumulation can occur on an ungrounded conductor, on the surface of an insulator, or in the body of an insulator.

When the charges accumulate on an ungrounded conductor, the ungrounded conductor can be considered to be a capacitor which holds the charge. One can think of this situation in terms of an idealized equivalent circuit where the charging cycle and the discharge cycle are independent. The charging of the capacitor from the constant voltage source follows an exponential form; and after the capacitor is charged and a path to ground is then connected, the discharge follows the inverse exponential form, Figure 2.4.

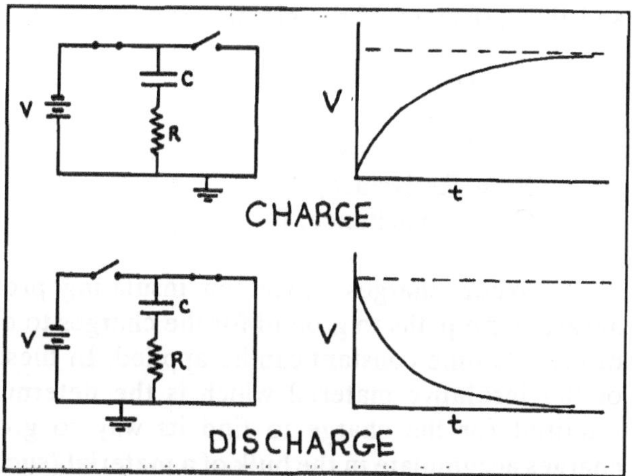

Figure 2.4: Charge and Discharge Circuits for Constant Voltage Source

The time it takes for the circuit to discharge will depend upon the resistance to ground and the size of the capacitor which holds the charge. The rate of decay is exponential.

$$Q = Q_o \exp(-t/\tau) \tag{8}$$

$Q \equiv$ Charge at time, t, C
$Q_o \equiv$ Initial charge, C
$t \equiv$ Time, s
$\tau \equiv$ Time constant, s

Table 2.1: Charge Remaining after Exponential Decay

Time Constants	Charge Remaining
1	0.368
2	0.135
3	0.050
4	0.018
5	0.007

The *time constant*, τ, is the time required for the charge to dissipate to 0.368 (1/e) of its original charge. It is also termed the *relaxation time* of the circuit. For all intents and purposes charges are considered to have practically dissipated after three time constants and completely dissipated after five time constants, Table 2.2.

For the discharge of a capacitor through a resistor to ground, the time constant is a function of the capacitance and the resistance of the circuit.

$$\tau = RC \tag{9}$$

$\tau \equiv$ Time constant, s
$R \equiv$ Resistance, Ω
$C \equiv$ Capacitance, F

When charges collect on insulating process materials there are always some paths to ground for the charges to dissipate and the principle of the RC time constant can be applied. In these cases it is the resistivity of the insulative material which is the determining factor for the time required for the charge to find its way to ground. When electrostatic charges accumulate in the bulk of a material (such as in an insulative liquid in a grounded tank or an insulative solid in a grounded silo) the relaxation time of the charge is determined by the materials resistivity, or conversely its conductivity.

$$\tau = \rho\epsilon\epsilon_o = \epsilon\epsilon_o/\kappa \tag{10}$$

$\tau \equiv$ Relaxation time. s
$\rho \equiv$ Resistivity, $\Omega{\cdot}m$
$\kappa \equiv$ Conductivity, S/m
$\epsilon\epsilon_o \equiv$ Permittivity, s/$\Omega{\cdot}m$

2.9 Host Material

As soon as a charge is accumulated the forces of mutual repulsion between the ions which make up the charge act to dissipate the charge. If the charge is in a conductive medium, the charge will easily dissipate through the medium to ground; but if the medium is a poor conductor, the charge cannot easily dissipate by finding its way to ground. The conductivity of the medium in which the charge is accumulated is very important in determining whether or not significant charges will accumulate. Again, there are two competing rates: the rate of charge separation and the rate of charge dissipation. When the rate of charge separation exceeds that of the rate of charge dissipation, electrostatic charges will accumulate. The common process examples of this are surfaces, powders, and liquids. The rate at which charges will dissipate through these will depend upon the surface conductivity, apparent conductivity, and volume conductivity respectively.

Also, as soon as a charge is accumulated a voltage difference is created with its concomitant electric field. This field resists the further separation of charge by exerting a counteracting force on the charges being separated. As charges continue to be separated the counteracting force increases until the energy being put into the system no longer overcomes the electrostatic forces already present, in which case a maximum is reached. This is generally true in industrial processes, but there are special cases, which will be developed later, where there are little or no counteracting forces.

All materials have some finite conductivity. There is no such thing as a perfect nonconductor and perhaps the term "nonconductor" is a misnomer which should not be used. On the other hand general usage has it that the term "nonconductor" refers to that electrical property of a material which does not conduct an electric current to any significant degree. So given that all materials have some conductivity, then a charge which has been accumulated on a insulative material will eventually dissipate and find its way to ground or its equal and opposite counterpart. The period of time for which the charge is retained on a material is characterized by the relaxation time of the material; the lower the conductivity, the longer the relaxation time

If a material has a comparatively high conductivity, charges can dissipate rather quickly such that no charges are accumulated; provided there is a conductive path to ground. But, if the material is insulated from ground by means of a non-conductor, the charge can be accumulated on the conductive material. The relaxation time for the dissipation of the charge will then be dependent upon the conductivity of the insulator.

2.9.1 Bulk Conductivity

Bulk conductivity has to do with the conduction of an electrostatic charge through the bulk of a material. When the bulk is a homogeneous solid or liquid, the bulk conductivity is the volume conductivity as given by Equation 3, and this can be viewed as the migration of unit charges from molecule to molecule through the interior of the material.

2.9.2 Surface Conductivity

The migration of charge across a surface of a material is characterized by its surface conductivity. This can be viewed as the migration of unit charges from one surface site to another. A solid will have a large number of extraneous molecules adsorbed on its surface and these adsorbed molecules will have a tremendous effect on the ease at which unit charges can migrate across the surface.

The most common example of this effect is the moisture which is adsorbed on the surface of an insulator. During periods of low humidity there is much less moisture adsorbed on a surface and the corresponding relaxation time is much longer. Therefore, during periods of low humidity, some electrostatic problems arise which are not present during periods of high(er) humidity, cf., ¶ 6.4.2.

2.9.3 Apparent Conductivity

When the conductivity of a pile of powder is in question, it is the apparent conductivity of the powder which is of interest; i.e., the rate at which a charge will relax from a pile of material. Sometimes it is the bulk conductivity of the material which is the controlling factor, but it is usually the surface conductivity of the granules which is the controlling factor. Apparent conductivities are determined experimentally and one must be careful to obtain data on the same material to which the data is to be applied. Apparent conductivity will vary with particle size, particle size distribution, humidity, and container geometry (Jones and Chan, 1989) - sometimes significantly.

The conductivity values given in the handbooks are those for homogeneous materials and have very little to do with the apparent conductivities if heaps of powders. One should not make the mistake of inferring an apparent conductivity from a volume conductivity.

It should be noted that when only "conductivity" is specified, it is implicit that it is the "volume conductivity" of Equation 3. When it is not, the terms "surface conductivity" or "apparent conductivity" should be used.

2.10 Separation vs. Relaxation

In practical situations, as charges are being separated they are simultaneously finding their way to ground. One must keep these two competing mechanisms in mind when analyzing electrostatic situations.

2.10.1 Constant Voltage Case

In the case where a constant voltage source is placed across a capacitor the charge at any time, t, will increase exponentially with time. Likewise, if there is a charged capacitor and there is then connected a path to ground, the amount of charge on the capacitor will decrease exponentially with time, Figure 2.4. In industrial processes it is unusual to have the constant voltage source where electrostatic potentials are concerned; it is more often a constant amperage source.

2.10.2 Constant Amperage Case

Consider a system which has a constant current generator (e.g., a streaming current) and a place (a capacitor) to accumulate the charge. Further consider that there is a resistance path to ground through which the charge may leak, but there are also places where incendive sparking to ground may occur. In analyzing such a situation to evaluate the probability of having spark discharge, the concept of an equivalent circuit can be used, Figure 2.5.

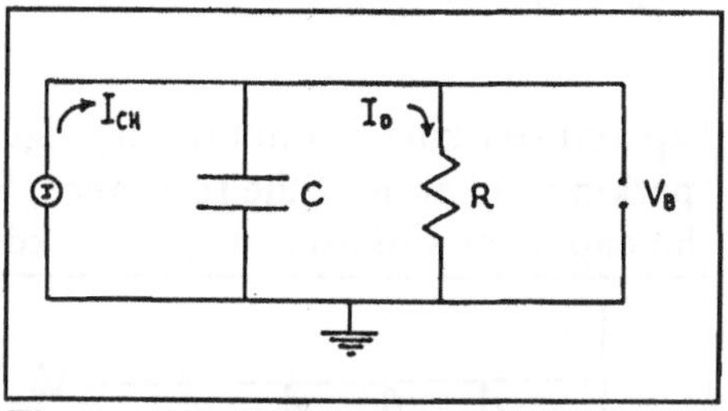

Figure 2.5: Equivalent Circuit for Simultaneous Charging and Discharging, Constant Amperage

In this circuit the capacitor, **C**, is charged with the charging current, $\mathbf{I_{ch}}$, and is concurrently discharged through the resistance, **R**, with a discharge current, $\mathbf{I_d}$. The voltage across the capacitor rises until the charging current equals the discharging current; therefore the voltage at equilibrium can be obtained from Ohm's Law; i.e., Equation 1.

$$V = I_{ch}R = I_dR$$

$V \equiv$ Voltage across the capacitor, V
$R \equiv$ Resistance, Ω
$I_{ch} \equiv$ Charging current from the source, A
$I_d \equiv$ Discharge current to ground, A

If the resistance, **R**, is increased beyond some critical value, then the voltage across the capacitor and the spark gap will exceed the breakdown potential, V_b, for the air in the spark gap and discharge across the gap will occur. When breakdown occurs the resistance in the discharge channel is

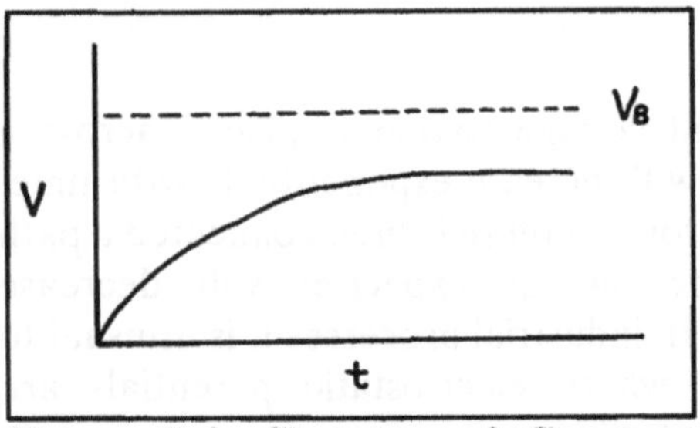

Figure 2.6: Charging of Constant Amperage Circuit, "Low" Resistance

very low and essentially all of the electrostatic energy in the capacitor will discharge across the gap and the energy in the discharge can be ascertained from the circuit parameters.

For a constant input current, if the resistance, **R**, in the circuit is "low", the voltage on the capacitor, **V**, will not exceed the breakdown voltage of the spark gap and sparking will not occur, Figure 2.6. On the other hand, repetitive sparking will occur if the resistance, **R**, is "high", because the voltage, **V**, on the capacitor will exceed the breakdown voltage, V_b, Figure 2.7.

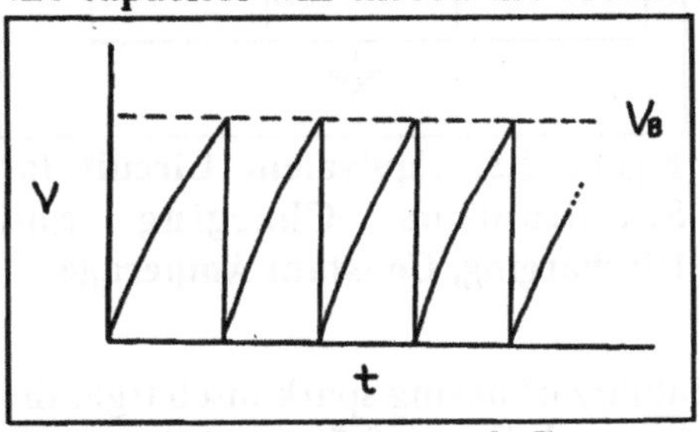

Figure 2.7: Discharge of Constant Amperage Circuit, "High" Resistance

When charges are separated in a system at a rate which is greater than the rate at which they dissipate, either a breakdown voltage, V_b, or a breakdown field strength, E_b, will be exceeded and discharge of some sort will occur. The various types of discharge are developed in Chapter 3.

2.11 Induction

Charging by induction occurs when a conductor is placed in an electric field. The electric field can be from an electrostatic charge being held on an insulator, or a space charge. This mechanism is quite common in industrial accident scenarios involving incendive electrostatic discharge.

Consider a block of a dielectric material which has an electrostatic charge bound to its top surface, Figure 2.8. If the dielectric is considered in this example to be a perfect insulator, then the surface charge is a *bound charge* and cannot be released into a spark. It must be remembered however, that there is an electric field associated with the charge bound on the surface of the insulator. If a block of metal (or any conductor) is brought into the electric field then charge separation will occur in the metal block by induction and result in a charge separation on the conductor, Figure 2.9.

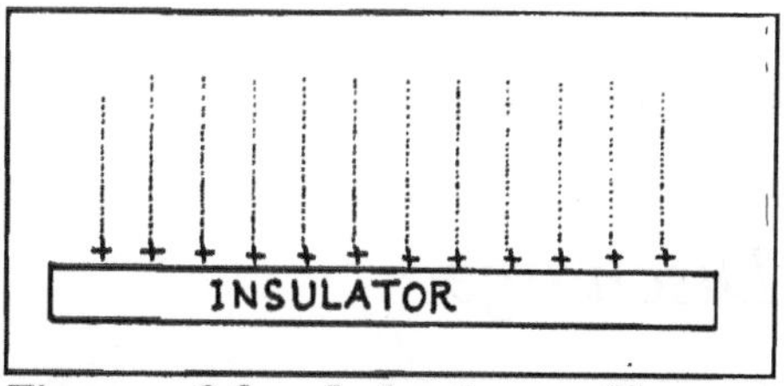

Figure 2.8: Induction, Charged Insulator

In this case, the positive charge on the insulator has induced charge separation on the metal block such that there is a negative *bound charge* on the bottom and an equal and opposite positive *free charge* on the top. The negative charge is bound on the bottom of the conductive block by the bound charge on the insulative block and will remain so as long as the blocks are not moved; however, the positive charge on the top is free to move. If a grounded electrode is touched to the top of the metal block, a spark discharge will occur. By this mechanism, a charge bound to a insulator has resulted in a spark discharge, Figure 2.10.

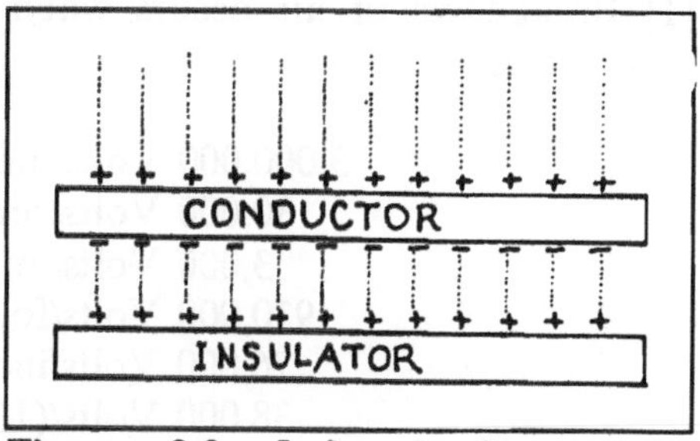

Figure 2.9: Induced Charge on Conductor

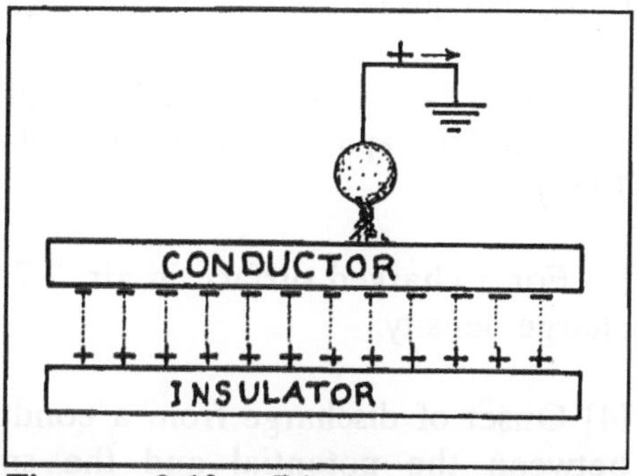

Figure 2.10: Discharge of Free Charge from Conductor

To go one step further, if the metal block is moved out of the electric field of the insulator along with its remaining negative charge, the negative charge remaining on the metal block is no longer bound and is free to move. This negative charge will therefore redistribute itself over the surface of the metal block and become a free charge. If a grounded electrode is again touched to the metal block, another spark will occur. Note that all the while, the initial charge remained bound to the insulator.

The process of induction is not 100% efficient in that the induced charge is always somewhat less than the initial charge.

Chapter 3

Discharge

Nutshells:

[1] Breakdown of air occurs when gradients equal or exceed 3×10^6 volts/meter.

3,000,000 Volts/meter
30,000 Volts/centimeter
3,000 Volts/millimeter
910,000 Volts/foot
76,000 Volts/inch
38,000 Volts/(1/2)inch

CAVEAT: Spacing will not exceed these values, but in inhomogeneous fields spacing may be less than those indicated above. (Haase, 1977)

[2] About 14,000 volts is needed to produce a spark between needle points one-half inch apart; over 20,000 volts is needed between spheres. (Eichel, 1967)

[3] For a charged surface in air, 2.7×10^{-5} C/m^2 is the maximum surface charge density.

[4] Onset of discharge from a conductor will begin as soon as the ratio between the potential and the smallest radius of curvature of the conductive surface reaches a value of 3×10^6 V/m. (Lüttgens and Glor, 1989)

3.1 Classification of Discharges

The classification of electrostatic discharges is empirical and phenomenological in nature and they may be classified into six types:

1. Corona Discharge
2. Brush Discharge
3. Bulking Brush Discharge (Cone Discharge)
4. Propagating Brush Discharge
5. Spark or Capacitor Discharge
6. Lightning

Whether a discharge is incendive to an ignitable mixture or not will depend upon the number of molecular fragments created in the space and time of the discharge. The general trend is that the more energetic the discharge, the more molecular fragments are formed and thus the more probable is ignition. Thus there is the notion of a *minimum ignition energy*, MIE; a discharge energy below which ignition will not occur but above which ignition will occur. This notion is not as intellectually honest as one would like, but it serves the purpose and variations are beyond the scope of the present discussions.

The energy in a given discharge can in principle be calculated based on the difference in electric fields before and after the event. The energy in two-electrode discharges can usually be calculated, but the energy in one-electrode discharges, e.g. corona and brush discharges, cannot be easily calculated, if at all.

3.2 Characteristics of Discharges

These discharges are separated into the different types by the character of their ionization of air when electrostatic energy is released.

3.2.1 Corona Discharge

When a charge is accumulated on a surface, there will be an electric field above the surface and the potential gradient in the field will be a maximum at the surface, Figure 3.1. That is, the strength of the electric field is the greatest at the surface of the material. The field strength at the surface is proportional to the surface charge density.

$$\sigma = \epsilon\epsilon_0 E \qquad (11)$$

$\sigma \equiv$ Surface charge density, C/m^2
$\epsilon\epsilon_0 \equiv$ Permittivity, F/m
$E \equiv$ Field strength, V/m

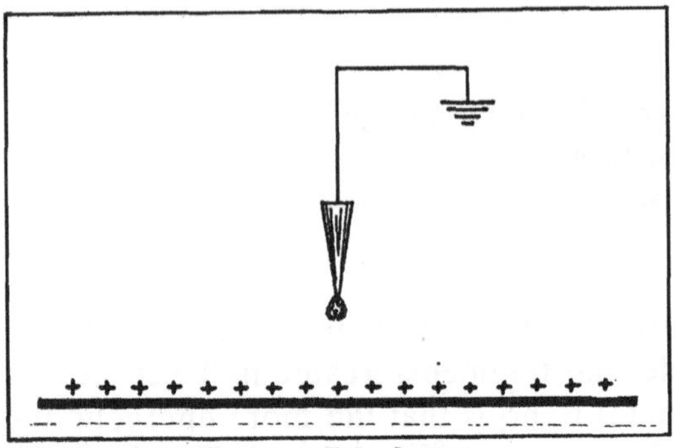

Figure 3.1: Corona Discharge

In air, as the surface charge density is increased the field strength at the surface increases. This process can continue until the breakdown strength of air is reached, at which time ionization of the air will occur. When the air becomes ionized the unlike ions will be attracted to the surface and the like ions will be repelled and a current will flow through the air. This causes the surface charge density and the field to drop and the current ceases, unless the charge is continually restored by the process. If the surface is a conductor and the charge is maintained by a power supply, the ionization and current flow is continuous. The ionization of the air above the surface is accompanied by a faint luminosity as the surface charges flow through the air and dissipate as a low energy density discharge termed *corona discharge*. If the charge is not continually restored, the electric field will decrease and the corona discharge will cease.

In either case, the maximum surface charge density in air corresponds to the breakdown field strength of air, i.e., 3×10^{6} V/m.

$$\sigma_{max} = \epsilon\epsilon_0 E_{max}$$

$$\sigma_{max} = (1)(8.85 \times 10^{-12})(3 \times 10^{6}) = 2.7 \times 10^{-5}\ \text{C/m}^2 = 27\ \mu\text{C/m}^2$$

Charge densities on free surfaces cannot effectively exceed this value because the charges will dissipate through the air in the corona discharge.

When a sharp point is inserted into an electric field in air, Figure 1.8 and Figure 3.1, the field is distorted, and the field strength is very high near the point since the field strength is inversely proportional to the radius of the point. When the point is sharp enough, corona discharge begins and charges diffuse into the surrounding air. Notice that the maximum field strength is exceeded only in the vicinity of the point. This is where the ions are created which then diffuse through the air to neutralize the surface charges. This is the principle of the tinsel bars used to remove "static" from webs of paper, fabric, and plastic. In the theoretical limit, points of zero radii would be capable of dissipating all of a surface charge when wiped over a charged surface.

Corona discharge is between a grounded electrode and a space where there is an electric field and is thus a one electrode discharge where the ions diffuse through the air to find their opposite signed counterparts. And since it is diffuse, there is a low concentration of molecular fragments in the stream and ignition of ordinary gasses, vapors, dusts, and mists does not occur. CAUTION: There are exceptions, see ¶ 5.6. This principle is used for charge dissipation in some industrial situations.

3.2.2 Brush Discharge

As discussed in the previous section, ionization of air occurs at a sharp electrode in an electric field. As the radius of the electrode increases, the character of the discharge changes. As the radius increases, a less diffuse discharge channel begins to form. This discharge channel begins to look like a brush and is termed *brush discharge* in contrast to corona discharge.

When a grounded, conductive, curved electrode is brought into an strong electric field or conversely when a strong electric field is created around a grounded, conductive electrode; the field is distorted, Figure 1.7, and brush discharge occurs at the surface of the electrode. Brush discharge is a one electrode discharge between a insulating charged part or surface (bags, pipes, walls, mists, dust clouds, bulked powders) and a conducting, **grounded** part (tools, vessel protrusions, instruments, fingers, etc.). A prerequisite for this type of discharge is a high field strength and a single electrode having a curved geometry, Figure 3.2.

It is basically unimportant how the electric field is generated as long as it is strong enough to result in the breakdown of the atmosphere at the surface of the electrode. Fields can be generated by charged insulator surfaces, charged bodies of non-conductive liquids, or space charged clouds of mists or aerosols.

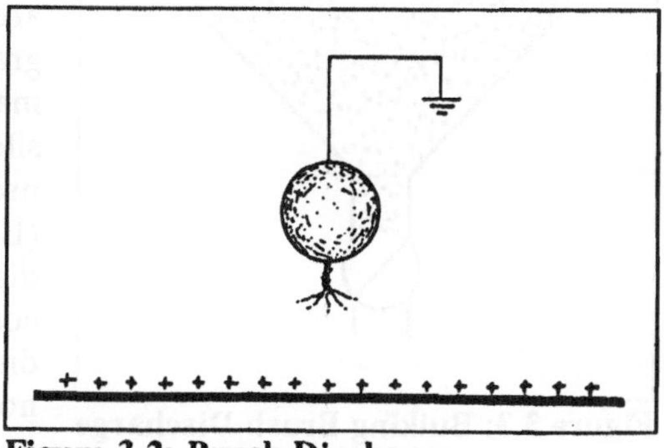

Figure 3.2: Brush Discharge

In contrast to a spark discharge, a corona or brush discharge does not lead to a discharge channel between two electrodes but leads to a diffuse discharge which issues forth from the site of the highest field strength at the surface of the electrode, and ends somewhere in space as a result of the decrease of the field strength with the distance from the electrode. In older literature brush discharges are referred to as "pre-breakdown streamers" and "insulator discharges".

For a given electric field, the character of a discharge at an electrode changes from a corona discharge to a brush discharge as one goes from a sharp electrode to a blunt electrode. In so doing the incendivity of the discharge increases with an increase in the radius of the electrode. The more sensitive the vapor to ignition, the less curvature of the electrode is required to effect ignition for a given electric field.

Incendive brush discharges are not restricted to spherical electrodes. They can occur at tips of fingers, straight or bent pipes, pipe bends, cables, casing edges, rivet heads, rims, or other conductive objects which do not have sharp points or crisp edges.

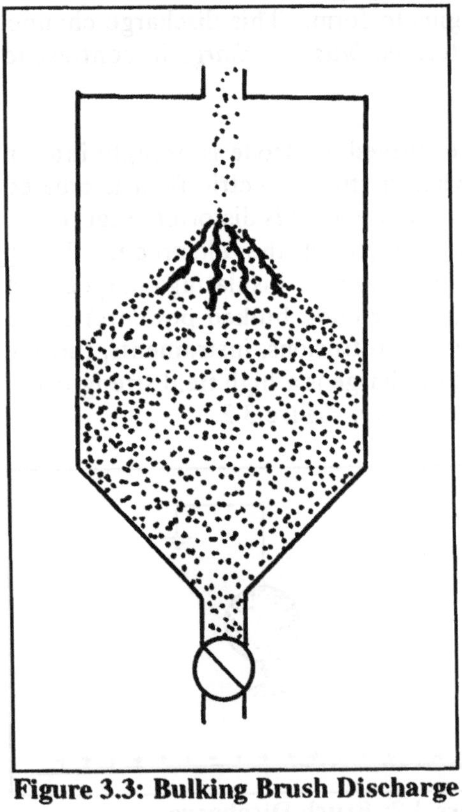

Figure 3.3: Bulking Brush Discharge

3.2.3 Bulking Brush Discharge

Bulking brush discharge is the type of discharge observed on the cone of a bulked heap of powder (thus, sometimes termed cone discharges). These discharges are associated with the transfer of granular polymeric insulating material into large containers and silos and appear different than the previously discussed discharges. (Bailey, 1987; Britton, 1988) These discharges are a direct result of the compaction of the powder as discussed in ¶ 5.4 and ¶ 6.3.2, and non-conducting organic powders are susceptible to this phenomenon, Figure 3.3.

3.2.4 Propagating Brush Discharge

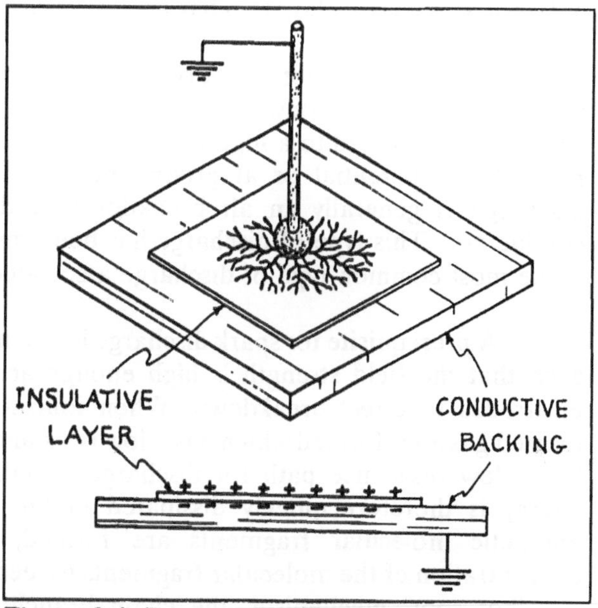

Figure 3.4: Propagating Brush Discharge

Propagating brush or Lichtenberg discharge is an energy-rich form of a brush discharge (Glor, 1987; Lüttgens, 1985; and Heidelberg, 1967). One condition for having this type of discharge is a highly charged insulating surface (eg., a film) backed with a grounded conductor. In such cases an extremely high charge density can be achieved on the surface of the insulator, perhaps an order of magnitude higher than that derived from Equation 17. This is because a large portion of the electric field is between the surface charge and the mirror charge induced on the metal backing and the breakdown field strength of the air is not exceeded. The majority of the field is between the layers of opposite charge and not in the air as would be the case if the metal backing were absent. Thus, a lot of energy can be stored in a metastable condition and when a brush discharge begins it propagates over the surface of the insulative layer, Figure 3.4.

Two conditions for having a propagating discharge are [1] a critical charge density of $\sim 2.5 \times 10^{-4}$ C/m^2 must be exceeded and [2] the thickness of the insulative layer must be less than 8 mm (Heidelberg, 1970). Higher charge densities or thinner insulative layers may lead to more energetic propagating brush discharges.

Several joules of energy can be released in a propagating brush discharge; therefore, in industrial practice ignition of any ignitable mixture should be expected.

3.2.5 Spark or Capacitor Discharge

Spark or capacitor discharge is the electrostatic discharge observed between two isolated conducting objects (people, products, and machines), one of them charged to a high potential and the second one charged to a much lower potential or at ground potential. The condition for the discharge is generally an air breakdown of the gap between the two conductors. This type of discharge has been investigated extensively and is the most common type of discharge associated with ignition hazards.

A prerequisite for spark discharge is a somewhat homogeneous field such that the field strength is high enough across the gap between the electrodes to effect breakdown. When ionization begins, a channel of ionized gasses is formed which has a low resistance. The stored charge then has a low resistance path for discharge, usually to ground. The stored energy is therefore quickly dissipated in the spark where all sorts of energetic molecular fragments are formed, Figure 2.5. When the concentration of the molecular fragments exceeds some critical value; i.e., the "hot spot" mechanism, the chain branching reactions of incipient combustion begin and ignition of the fuel/oxidizer mixture is accomplished.

The critical field strength (volts per meter) can be exceeded either by closing the gap or increasing the voltage or charge. Both mechanisms are common in industrial operations. A gap is closed when a person receives a shock when reaching for the door knob. There is a voltage and charge increase when a charged liquid is added to a metal bucket until a spark jumps to ground.

There is a correlation between the energy in the spark and ignition of explosives, flammable vapors, and dusts. The concept of minimum ignition energy is introduced as a useful criterion for ignition; if the energy stored in the process exceeds the minimum ignition energy required for the ignition of the process material, then the criterion has been exceeded and ignition is presumed.

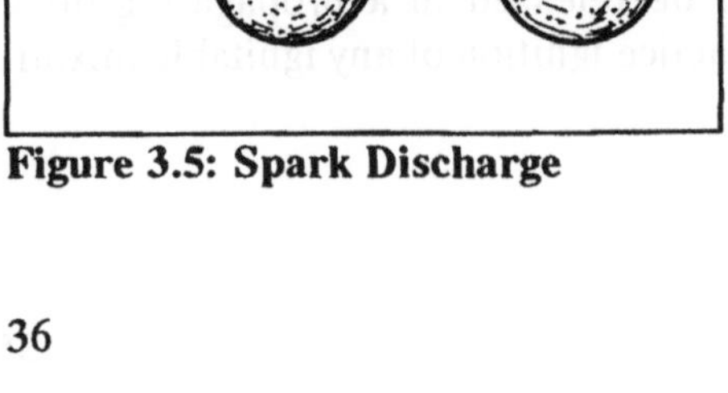

Figure 3.5: Spark Discharge

Spark discharge generally occurs between the capacitors found in the workplace; thus, they are usually equated to the purely capacitive discharges generated in the laboratory. Any ungrounded conductor in the workplace - cans, buckets, drums, tanks, carts, vehicles, machinery components, and humans to name a few - can constitute a capacitor to store a charge. When such an item is involved, the capacitance of the item and its potential or the amount of charge stored on it can be used to reckon a spark discharge energy by the use of the relationships given in Chapter 5.

For purely capacitive discharge this energy can then be compared to the minimum ignition energy for an assessment of potential hazard. Electrostatic sparks in industrial processes are usually many times more energetic than the published minimum ignition energies of vapors; therefore, the discovery of a scenario for spark discharge is usually enough to identify a hazard. It is unacceptable to continue the operation of a process where electrostatic sparks were known to occur around flammable atmospheres whatever their energy content may be.

3.2.6 Lightning

Under certain atmospheric conditions a tremendous amount of electrostatic energy can be stored in clouds. This energy can be released in the familiar lightning strike, Figure 3.6. There is no doubt that any ignitable mixture can be ignited by atmospheric lightning, but the question sometimes arises as to the possibility of lightning-like discharges within process equipment. It has been shown that lightning-like discharges do not occur in process equipment of volumes less than 60 m^3 nor in cylindrical containers with a diameter of less than 3 m, regardless of their heights (Boschung et. al., 1977).

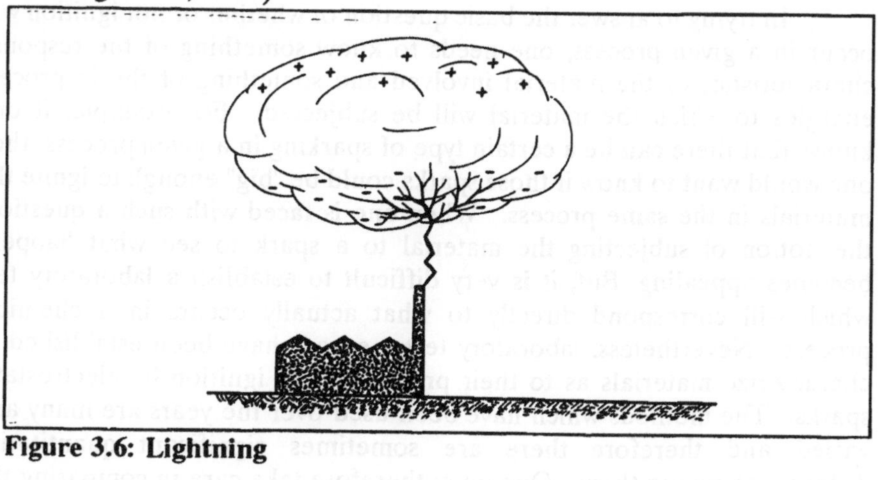

Figure 3.6: Lightning

Chapter 4

Minimum Ignition Energies

Nutshells:

[1] The concept of a minimum ignition energy for ignition applies only to capacitive spark discharge.

[2] The minimum ignition energy of a spark is a function of the electrode spacing and the optimum spacing is approximately equal to the quenching distance, which is of the order of 2 mm for most hydrocarbons. (Eichel, 1967)

[3] The minimum ignition energy for ordinary vapors is considered to be 0.25 millijoules. (API RP2003, 1991; NFPA 77, 1988).
Caveat: There are exceptions; among them are hydrogen, acetylene, and carbon disulfide. For these materials 0.01 millijoule sparks should be considered as being incendive.

4.1 Testing of Materials

In trying to answer the basic question of whether or not ignition will occur in a given process, one needs to know something of the response characteristics of the material involved and something of the in-process energies to which the material will be subjected. For example, if one knows that there can be a certain type of sparking in a given process, then one would want to know if those sparks could be "big" enough to ignite the materials in the same process. When one is faced with such a question, the notion of subjecting the material to a spark to see what happens becomes appealing. But, it is very difficult to establish a laboratory test which will correspond directly to what actually occurs in a chemical process. Nevertheless, laboratory test methods have been established to characterize materials as to their propensity for ignition by electrostatic sparks. The methods which have been used over the years are many and varied and therefore there are sometimes significant quantitative differences among them. One must therefore take care in comparing the data from one laboratory with that of another.

4.2 Minimum Ignition Energy, MIE

The mechanism whereby a fuel/oxidizer mixture is ignited is that of forming a critical concentration of molecular fragments, whether it be by electrostatic or other means. There is both a temporal and a spatial critical concentration required to start the chain branching combustion reaction; i.e., a "hot spot". The exact mechanism for a given mixture and given conditions is a very complicated one and in borderline cases there are few if any guidelines one can rely upon to predict whether or not ignition will occur from a given set of circumstances. It is intuitively obvious that energy is expended in the creation of the required molecular fragments; therefore, it follows that the higher the energy input, the more likely the ignition and an energy term should be used to characterize the incendivity of an electrostatic discharge. This leads to the notion of a *minimum ignition energy*, MIE, for ignition; i.e., an energy below which the ignition of an explosible mixture will not be effected and above which ignition can be effected.

The minimum ignition energy requirement for a combustible substance can be important in the assessment of hazards in a plant. It can serve as a guideline for determining the scope of the protective measures to be taken and to provide an insight for the general understanding of ignitions by static electricity. The minimum ignition energy of a combustible substance is the lowest electrical energy stored in a capacitor which, when discharged, will ignite the material at the conditions of the test..

Minimum ignition energies (MIEs) are used to characterize the incendivity of capacitive spark discharges. Investigators strive to vary the experimental conditions so that a minimum is indeed achieved. Then it is assumed that if the minimum can be exceeded in a process, ignition should be expected. Implicit in the notion of an MIE is that there is a sharp distinction between ignition and nonignition for a given energy input; i.e., there is a critical value below which ignition will never occur and above which ignition will always occur. Such is not the case because there is a regime where ignition occurs only some of the time; a probabilistic cumulative distribution. Experimenters usually strive to determine a threshold initiation energy when determining and reporting an MIE.

It should be pointed out that the spark discharge energies commonly found in industrial processes from ungrounded equipment are usually many times that required for the ignition of an ideal mixture under ideal conditions. In such cases MIEs are only of academic value. On the other hand MIEs are used as effective qualitative guideposts for the characterization of hazards in chemical processes.

If one is to use MIEs one should have an experimental technique which relates to what is actually going on in the chemical process. At the present time progress is being made to develop a standard test so that a body of homogeneous data can be established. The data which is in the present literature comes from many sources and therefore from many different test systems and it is to be expected that data from one laboratory will differ significantly from data taken at another laboratory. Data from different laboratories can sometimes differ by as much as an order of magnitude for the same substance so a bit of caution is warranted in the use of the data. Data from the same laboratory can perhaps be used for qualitative comparisons; i.e., is one material more sensitive to ignition than another.

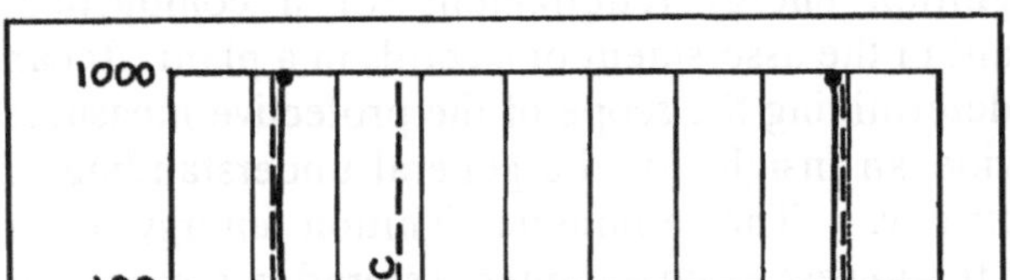

Figure 4.1: MIE as a Function of Benzene Concentration (Britton, 1992)

4.2.1 MIEs of Gasses and Vapors

The MIE for a gas or vapor varies with the stoichiometry of the mixture and there is an optimum fuel/air mixture which will give the lowest MIE. At or near the lower flammable limit, LFL, or the upper flammable limit, UFL, the spark discharge energy required to effect ignition may be a joule or so. As the mixture approaches the stoichiometric point (the concentration of fuel in air where the products of combustion are carbon

Table 4.1: LMIEs of Selected Gasses and Vapors

Fuel	LMIE, mJ	Fuel	LMIE, mJ
Acetaldehyde	0.37	Hydrogen	0.016 @ 28%
Acetone	1.15 @ 4.5%	Hydrogen sulfide	0.068
Acetylene	0.01 @ 8.5%	Isooctane	1.35
Acrolein	0.13	Isopentane	0.21 @ 3.8%
Acrylonitrile	0.16 @ 9.0%	Isopropyl alcohol	0.65
Allyl chloride	0.77	Isopropyl chloride	1.08
Benzene	0.2 @ 4.7%	Isopropyl amine	2.0
1,3-Butadiene	0.13 @ 5.2%	Isopropyl mercaptan	0.53
Butane	0.25 @ 4.7%	Methane	0.21 @ 8.5%
n-Butyl chloride	1.24	Methanol	0.14 @ 14.7%
Carbon disulfide	0.009 @ 7.8%	Methylacetylene	0.11 @ 6.5%
Cyclohexane	0.22 @ 3.8%	Methyl ethyl ketone	0.53 @ 5.3%
Cyclopentadiene	0.67	Methyl butane	0.25
Cyclopentane	0.54	Methyl cyclohexane	0.27 @ 3.5%
Cyclopropane	0.17 @ 6.3%	Methyl formate	0.4
Diethyl ether	0.19 @ 5.1%	n-Pentane	0.28 @ 3.3%
Dihydropyran	0.36	2-Pentane	0.18 @ 4.4%
Diisobutylene	0.96	Propane	0.25 @ 5.2%
Diisopropyl ether	1.14	Propionaldehyde	0.32
Dimethyl ether	0.29	n-Propyl chloride	1.08
Dimethyl sulfide	0.48	Propylene	0.28
Di-t-butyl peroxide	0.41	Propylene oxide	0.13 @ 7.5%
Ethane	0.24 @ 2.5%	Tetrahydrofuran	0.54
Ethyl acetate	0.46 @ 5.2%	Tetrahydropyran	0.22 @ 4.7%
Ethylamine	2.4	Thiophene	0.39
Ethylene	0.07	Toluene	0.24 @ 4.1%
Ethylene oxide	0.065 @ 10.8%	Tetraethylamine	0.75
Furan	0.22	Vinyl acetate	0.7
Heptane	0.24 @ 3.4%	Vinyl acetylene	0.082
Hexane	0.24 @ 3.8%	Xylene	0.2

Reference: Britton, 1992 (selected data)

Table 4.2: LMIEs[a] of Selected Hydrocarbons at Reduced Pressures.[b]

Pressure, atm.	LMIE, mJ		
	Methane	Ethane	Propane
1	0.45	0.24	0.26
0.50	1.4	0.66	0.70
0.33	2.6	1.8	1.8
0.25	7.0	4.5	5.5
0.10	26	-	-

[a] At ambient temperatures of ~25° C
In an atmosphere of $O_2/(O_2 + N_2) = 0.21$
[b] Inferred from graphs of Lewis and von Elbe, 1961

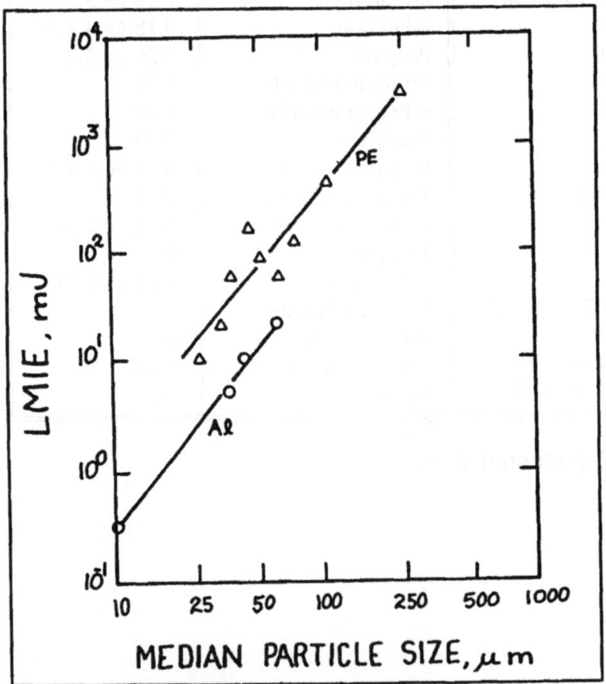

Figure 4.2: LMIE as a Function of Median Particle Size (Bartknecht, 1989)

Table 4.3: MIEs[a] of Selected Gasses and Vapors at 25° C and 150° C

Fuel	MIE, mJ 25° C	150° C
Acetone	0.406	0.188
Benzene	0.23	0.145
Cyclohexane	0.24	0.145
Diethyl Ether	0.25	0.089
Dimethyl Ether	0.345	0.22
Ethane	0.292	0.208
Ethyl Acetate	0.43	0.218
Ethylene	0.121	0.062
Ethanol	0.40	0.13
Heptane	0.26	0.082
Hydrogen	0.011	0.0051
Isobutylene	0.471	0.246
Isobutane	0.376	0.282
Methane	0.30	0.167
Propane	0.476	0.265
Propylene	0.24	0.187
Toluene	0.26	0.106

[a] In air at ambient pressure, stoichiometry unspecified.
Reference: Gavrilenko, 1973

dioxide and water) the MIE becomes much less for a given mixture, Figure 4.1. The optimum concentration which gives the lowest MIE is not the stoichiometric mixture but a mixture which is a bit fuel rich. At this richer mixture a typical MIE for an ordinary vapor is usually less than one millijoule. The energy required to ignite the optimum mixture has been termed the *lowest minimum ignition energy*, LMIE, (eg., Britton, 1992) and is the one which is now being given in the tables, Table 4.1.

Some of the older literature (eg., Calcote et al., 1952) give MIEs for stoichiometric mixtures and thus should not be confused with the more conservative LMIEs. Many of the tables in the literature contain MIEs from various sources and again a bit of caution is warranted in the use of the data.

Table 4.4: MIEs[a] of Selected Fuels in Air and Oxygen Atmospheres

Fuel	MIE, mJ.		
	Air	Oxygen	Ratio
Acetone	1.15	0.0024	479
Acetylene	0.017	0.0002	85
n-Butane	0.25	0.009	28
Cyclopropane	0.18	0.001	180
Diethyl ether	0.20	0.0013	153
Ethane	0.25	0.002	125
Ethylene	0.07	0.001	70
n-Hexane	0.228	0.006	48
Hydrogen	0.017	0.0012	14
Methane	0.30	0.003	100
Coal	60	10	6

[a] Nominal 1 atm., 25° C, stoichiometry unspecified.
References: NFPA, *Fire Protection Handbook*, 1986, and BuMines RI 6597

One must recognize that MIEs vary with most everything; therefore, if one has a chemical system at different pressures, temperatures, or atmospheres, one should make some account for the variability of the MIE for those conditions.

- The MIE of a gaseous mixture varies inversely with pressure. A decrease in pressure results in an increase in the MIE, Table 4.2.

- The MIE of a given gaseous mixture varies inversely with temperature. An increase in temperature will result in a decrease in the MIE, Table 4.3.

- If the supporting atmosphere is something other than air, there can be a significant difference in the MIE. For instance, an increase in oxygen concentration will result in a decrease in the MIE of a given fuel, Table 4.4.

The general rule of thumb for the LMIE of hydrocarbon gases is

0.25 mJ (API RP2003, NFPA 77), and this works reasonably well for other ordinary vapors. But, one must be acutely aware of some exceptions:

- Strained molecules (e.g., acetylene, ethylene, and ethylene oxide) have much lower LMIEs.
- Easily ignitable gasses (e.g., hydrogen and hydrogen sulfide) have lower LMIEs
- Some unusual materials (e.g., carbon disulfide which is almost pyrophoric) have low LMIEs.

4.2.2 MIEs of Dusts

The determination of MIEs in dusts is more complicated than that for gasses and vapors. Particle size, particle shape, particle concentration, turbulence, ignition delay time, moisture content, electrode shape, electrode spacing, circuit inductance, circuit resistance, supporting atmosphere, temperature, and pressure are among the parameters which have a measurable effect upon the determination of an MIE (See Bartknecht, 1989). But again the notion of LMIE as developed for gasses and vapors is applicable to dusts. If all of the experimental conditions are tailored to yield the lowest MIE, then this LMIE can be taken to represent the most sensitive combination of in-process conditions for the given material.

Over the years The United States Department of the Interior, Bureau of Mines performed dust explosibility tests on a myriad of substances in the Hartmann apparatus. These data have been reported in several compendia, but most of the data are of little value because they apply only a dust from a specific operation; eg., a scraping of a powder from a duct. However, the Bureau of Mines reports do contain some data for generic materials which is of general applicability. A selection of these data is given in Table 4.5.

Other investigators have determined MIEs and LMIEs in their own equipment and obtained results which are sometimes quite similar to those of the Bureau of Mines, but are sometimes quite different. Differences of an order of magnitude have been observed. In order to account for such differences, the notion of having a standard dust to which all of the various

Table 4.5: MIEs[a] of Selected Dusts as Reported by the Bureau of Mines

Dust	MIE, J	Ref.[b]
Aluminum, flake	0.100	RI 6516
Aluminum, atomized	0.050	RI 6516
Benzoic acid	0.020	RI 7132
Cellulose acetate	0.015	RI 5971
Charcoal	0.020	RI 6597
Cinnamon	0.03	RI 5753
Coal (Pittsburgh)	0.060	RI 6597
Cornstarch	0.03	RI 5753
Dextrin	0.04	RI 7208
Ethyl cellulose	0.01	RI 5971
Lycopodium	0.04	RI 5753
Nitrostarch	0.04	RI 7208
Pea flour	0.04	RI 5753
Polycarbonate	0.025	RI 5971
Polyethylene	0.010	RI 5971
Polymethylmethacrylate	0.015	RI 5971
Polypropylene	0.030	RI 5971
Polystyrene	0.040	RI 5971
Polyvinyl acetate	0.16	RI 5971
Soap powder	0.10	RI 7208
Stearic acid	0.02	RI 7132
Sugar, powdered	0.03	RI 5753
Trinitrotoluene	0.075	RI 7208
Wheat flour	0.05	RI 5753

[a] MIEs as performed in the Bureau of Mines Hartmann apparatus.

[b] Bureau of Mines Report Number

data could be related has been employed. The Bureau of Mines related their data to Pittsburgh coal dust as the baseline; therefore many of the other investigators related their data to it as well. Other investigators simply established their own body of homogeneous data within which a particular dust could be related. At present (1997) there is an effort to establish lycopodium as the baseline material. Lycopodium is attractive because the spores have a uniform particle size of some 30 μm and has dust explosibility characteristics which are typical of many industrial materials (Eckoff, 1991). There are two species of lycopodium and one must take care to use the reticulate *lycopodium clavatum* rather than the rugulose *lycopodium alpinum* (Thomas et.al., 1991) since it is the former which has been used as the standard dust.

As would be expected the LMIE of a particular dust varies with most everything. Many studies have been made on the variation of LMIE with the mean particle size of the dust. In general, the LMIE of a dust is a direct function of the mean particle size and the LMIE may vary by two orders of magnitude in going from a mean of 10 μm to a mean of 500 μm, Figure 4.2 (Bartknecht, 1989).

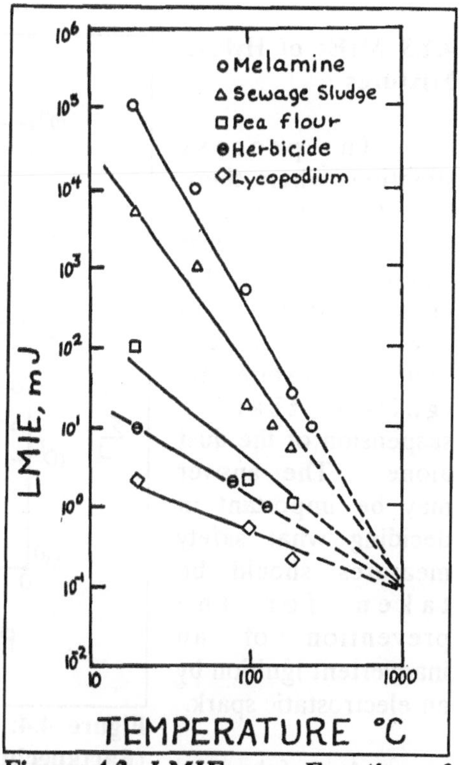

Figure 4.3: LMIE as a Function of Temperature, (Bartknecht, 1989)

A useful observation has been made for the variation of LMIE of a dust with temperature. It has been observed for several typical dusts that as temperature increases, the LMIE decreases, until (by extrapolation) an LMIE of 0.1 mJ is reached at a temperature of 1000° C, Figure 4.3 (Bartknecht, 1989). This observation is useful in estimating an LMIE of a dust at a temperature other than ambient by extrapolation; however, Bartknecht (1989) gives us the caveat that such an extrapolation is only valid for those dusts which have explosion indices independent of the applied ignition energy.

The question concerning the variation of LMIE of a dust with humidity seems to be quite prevalent and usually confusing. The humidity (water content) of the product is one thing while the humidity of the supporting atmosphere is quite another. The effect of water in the product has a very pronounced effect upon the LMIE of a dust, Figure 4.4 (Bartknecht, 1989). Therefore, determinations of ignition behavior should be performed on samples of dry dust. The effect of the humidity of the air used in the test is only secondary (or perhaps even tertiary) such that no special precautions need be taken in regard to the humidity of the laboratory in which the tests are being performed, other than to keep the powder dry.

4.2.3 MIEs of Hybrid Mixtures

In process situations where there is both a flammable vapor and a dust suspension, the question arises as to how much easier can such a mixture be ignited than a suspension of the dust alone. The answer may be important in deciding what safety measures should be taken for the prevention of an inadvertent ignition by an electrostatic spark.

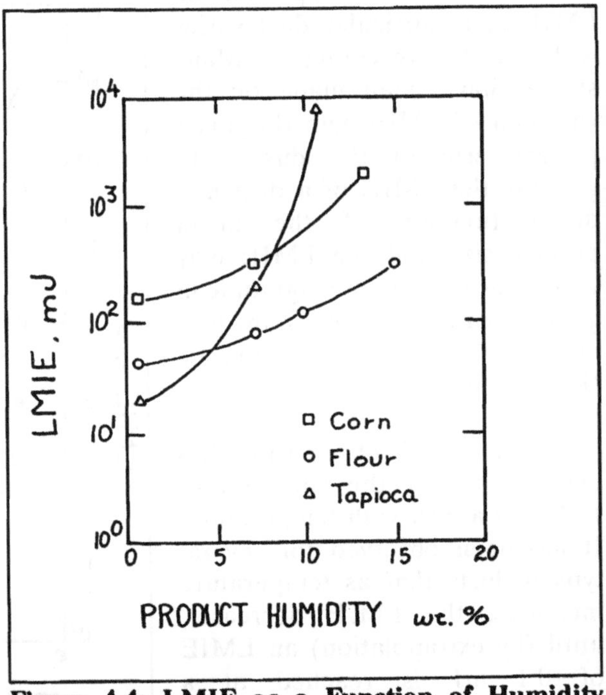

Figure 4.4: LMIE as a Function of Humidity (Bartknecht, 1989)

A useful model of such ignitions is reported by Bartknecht (1989) where the influence of propane content in the supporting atmosphere is shown for suspensions of five dusts having different LMIEs. On a semi-logarithmic plot there is a linear correlation between the LMIE of the dust suspension with propane concentration, and the lines intersect at the LMIE of propane, Figure 4.5 (Bartknecht, 1989). Thus if one has the LMIE of a dust and the LMIE of the vapor, one can estimate the LMIE of a hybrid mixture of the two by interpolation on a semi-logarithmic plot.

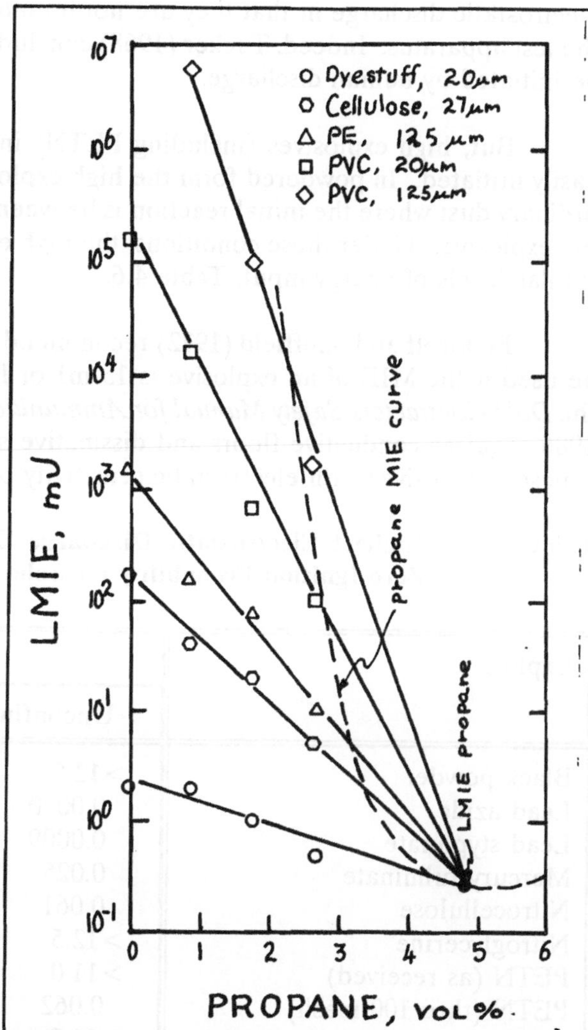

Figure 4.5: LMIE for Hybrid Mixtures (Bartknecht, 1989)

4.2.4 MIEs in Enriched Oxygen Atmospheres

The minimum ignition energies in enriched oxygen atmospheres can be many times less than those in air; for some materials the minimum ignition energy in pure oxygen can be 100 times less than the corresponding value in air, Table 4.4.

4.2.5 MIEs of Explosives

The nature of a spark discharge in a solid fuel/oxidizer mixture is that it creates a hot spot within the condensed phase material. For instance, if a charged person touches a high explosive, the electrostatic energy can discharge from his body into the body of the solid (or liquid) explosive. As this energy is dissipated, molecular fragments are created within the mass of the material. If the concentration of these fragments exceeds the critical and temporal value; i.e., a hot spot, then the explosive may be initiated.

The primary explosives by their very nature are rather sensitive to electrostatic ignition. On the other hand some of the high explosives in their cast or pressed form can be very insensitive to initiation by a direct electrostatic discharge in that they are not initiated at the upper limits of the test apparatus. Indeed, Tucker (1968) concluded that PETN would not be initiated by human discharge.

But, high explosives (including PETN) in powdered form can be easily initiated. In powdered form the high explosives behave more as an ordinary dust where the initial reaction is between the surrounding air and the explosive. Under these conditions, the high explosives are initiated at similar levels of energy input, Table 4.6.

Fedoroff and Sheffield (1972) recommend that dissipative footwear be used if the MIE of an explosive is 15 mJ or less. On the other hand, the *DoD Contractors Safety Manual for Ammunition and Explosives*, March 1986 requires conductive floors and dissipative shoes "at operations with exposed explosives with electrostatic sensitivity of 0.1 joule or less."

Table 4.6: Highest Electrostatic Discharge Energy at 5000 Volts for Zero Ignition Probability for Selected Explosives

Explosive	Energy, J	
	Unconfined	Confined
Black powder	>12.5	0.8
Lead azide	0.0070	0.0070
Lead styphnate	0.0009	0.0009
Mercury fulminate	0.025	0.025
Nitrocellulose	0.061	3.1
Nitroglycerine	>12.5	0.90
PETN (as received)	>11.0	0.21
PETN (thru 100 mesh)	0.062	0.21
Tetryl (as received)	>11.0	4.68
Tetryl (thru 100 mesh)	0.007	4.38
TNT (as received)	>11.0	4.68
TNT (thru 100 mesh)	0.062	4.38

Reference: Fedoroff and Sheffield, 1972

Chapter 5

Discharge Energies

Nutshells:

[1] Electrostatic sparks coming from ungrounded conductors are usually many times more energetic than the MIEs of vapors and in such cases the precise value of the MIE is seldom important in the evaluation of hazard. (Cross, 1987)

[2] Sparks from the human body are less incendive by a factor of 2 to 4 than purely capacitive sparks. (Berkey et. al., 1988)

[3] Incendive brush discharges can occur in grounded equipment if the geometry of the electrode and the strength of the electric field are favorable. (Heidelberg, 1967)

[4] Incendive discharges can occur on the surfaces of highly charged insulative materials. (Gibson, 1965; and Glor, 1981)

5.1 Ignitions by Electrostatic Discharges

There is no doubt that electrostatic sparks can ignite flammable mixtures; after all, this is the thrust of the entire text. It is a matter of determining the conditions where many things come together at the right place at the right time for ignition to be accomplished. In an explosible fuel/oxidizer mixture the creation of a critical concentration, in both space and time, of energetic molecular fragments will lead to a self propagating combustion reaction; i.e., the mixture will be ignited. In general, this critical concentration of molecular fragments is not exceeded in corona discharge; but when the geometry of the discharge electrodes changes from a sharp to a blunt configuration, the critical concentration of molecular fragments will increase and ignition of many flammable mixtures can be accomplished. Corona and brush discharges are one electrode discharges where there is a diffuse ionization region between a grounded electrode and the surrounding space. This is in contrast to classical spark discharge where an ionization channel develops between two conductive electrodes which are at a different potential. When enough energy is dumped into this ionization channel, ignition is accomplished. When the spark discharge is purely capacitive, some simple mathematical relationships can be applied but these relationships do not apply to corona, brush, or bulking brush discharges. In these latter instances the concept of even having a capacitance does not apply.

5.2 Capacitive Discharges

For a purely capacitive discharge, the energy stored in the capacitor can be reckoned from the following equations:

$$C = \frac{Q}{V} \tag{12}$$

$$W = \frac{1}{2}CV^2 \tag{13}$$

$$W = \frac{1}{2}QV \tag{14}$$

$$W = \frac{Q^2}{2C} \tag{15}$$

It should be noted that these relationships apply to the energy stored in the capacitor and not to the energy in the discharge; there are always some energy losses in the circuit which does not get into the spark. When there is no appreciable inductance or resistance in the discharge circuit, which is more often the case, then the capacitor relationships give a very good determination of the energy in the spark. That is to say that where a charged and ungrounded conductor (eg., a metal 55 gallon drum) discharges to a metallic ground, then the relationships are valid. They do not apply directly in the case of human discharge since there is a significant inductance and resistance in the human body.

5.2.1 Human Sparks

It is possible for the human body to store many times the electrostatic energy required for the ignition of gases, flammable vapors, dusts, and explosives. It has been shown that sparks from the human body with a given energy content are less incendive to flammable vapors than are purely capacitive sparks (Berkey et al., 1988). It generally requires from 2 to 4 times as much stored energy to effect ignition with human sparks because of the resistance and inductance in the circuit.

A common condition in human spark discharge scenarios is that the person is insulated from ground, either by standing on a insulating surface or by wearing insulating footwear. It is common practice in some industrial situations to provide employees with conductive footwear and conductive flooring.

For the purposes of estimating discharge energies the capacitance of a human is generally taken to be 200 pF (Haase 1977), but values ranging from 80 pF to 600 pF have been reported (Berkey et. al. 1988). The resistance of common footwear can range between a few hundred ohms to several teraohms, cf. Table 9.2; therefore relaxation times can vary from microseconds to minutes, depending on conditions.

The classic example of human discharge is the walking across an insulative carpet and reaching for the door knob. In this process charges are generated by the contact and separation of a person's shoes with the carpet. When the person picks up one foot, the charge on the shoe distributes itself over the person; and each time separation occurs, more charge is added. In this way the charges are accumulated on the person where they can subsequently discharge. (Brundrett, 1977)

Charges can also be induced on a person when the person is in an electrostatic field. For example, if there is a charge on a pile of dielectric material and a person with insulative shoes walks by the pile, a charge will be induced on him. If he touches a grounded object while in the field, a spark (which may be incendive) will occur. If he then walks out of the field, he will take the opposite charge with him which can likewise result in another spark which may also be incendive; cf., Induction, ¶ 2.11.

The range of energies in sparks from human discharges can range from microjoules to tens of millijoules. A one millijoule spark is perceptible, a ten millijoule spark is a prick, a 30 millijoule spark is a sharp prick, and a hundred millijoule spark results in a slight jerk (Klinkenberg, 1958). Undetectable human sparks can do damage to electronic circuits and may be incendive to some sensitive materials such as carbon disulfide. Any detectable human spark should be considered as being incendive to ordinary flammable vapors.

5.2.2 Clothing

Many times during an investigation of an incident, investigators will overly concern themselves with the type of clothing an operator was wearing. The concern stems from the notion that if the operator was wearing synthetic materials, ignition could have occurred from electrostatic discharge because of the clothing. Such scenarios for ignitions are not credible unless the operator had removed some clothing, a jacket for instance.

If a person is wearing two different articles of clothing, a wool jacket over a polyester shirt for example, normal movement will result in a charge separation at the wool/polyester interface; however, as long as the wool and the polyester remain close together, the charges - even though separated - will bind each other to the surfaces at the interface and the system will essentially be electrically neutral. In such cases the fabrics will cling together because of the electrostatic attraction between opposite charges. As long as the clothing remains on the person, there will be no net charge accumulated on the person's body which can be discharged. On the other hand if the person removes the jacket, the positive charge on the jacket will be separated from the negative charges on the polyester shirt (cf., Table 6.3), and this remaining negative charge will then be induced on or transferred to the conductive person. Then, the charge on the person can discharge in a spark which may be incendive to flammable vapors or dusts.

Likewise a person sitting in a chair can separate charges by moving his clothing against the material of the chair. The charges so separated will be bound to each other as long as the person remains seated so that the system is essentially electrically neutral. When the person gets up however, the charges will then be accumulated, and if the person then touches a grounded object, incendive discharge may occur - depending on conditions. The classic example of this scenario is the sliding across a car seat during periods of low humidity and then touching a grounded metal object.

5.3 Brush Discharges

In capacitive discharges the charge which has been accumulated on a conductor rapidly moves to ground through the plasma of the discharge channel. But, when the accumulated charge does not reside on a good conductor and cannot readily move into and through the plasma, a diffuse discharge results. The temptation here is to term such events as "frustrated discharges" since the stored energy has no low resistance path to ground through which essentially all of the charge can flow. Corona, brush, and bulking brush discharges fall into this category.

Since all of the stored energy is not dissipated in a single event, the notion of an MIE does not serve as a criteria of ignition for these diffuse discharges and the temptation to use MIEs as a measure of the ease of ignition should be resisted.

For a given electric field, the character of a discharge at an electrode changes from a corona discharge to a brush discharge as one goes from a sharp electrode to a blunt electrode. At a sharp electrode the discharge is a diffuse corona discharge where the critical concentration of molecular fragments is not usually achieved. As the radius of the electrode increases, the discharge begins to form a core. The length of this core increases as the radius of the electrode increases and the concentration of molecular fragments in the discharge increases (cf., Figure 3.2). When the critical concentration of molecular fragments is achieved, ignition occurs. Thus, the incendivity of the discharge increases with an increase in the radius of the electrode. The more sensitive the vapor to ignition, the less curvature of the electrode is required to effect ignition for a given electric field. The electric field can come from any configuration of accumulated charge, but there are two particular instances of interest: a space charge and a surface charge.

5.3.1 Brush Discharges in Spaces

There are instances where a space charge can accumulate in the presence of flammable atmospheres. Examples would include a charged liquid in a tank where the outage contained flammable vapors or a charged mist in an explosible atmosphere. In such cases it matters little how the electric field is created but the geometry of the grounded electrode matters a lot.

Brush discharge can occur at a grounded electrode [1] when the electrode is inserted into an electric field, [2] when the electrode is fixed and the electric field is created around it, or [3] when there is some of both. If a flammable atmosphere exists where the brush discharge takes place, ignition may result.

Ordinary hydrocarbon and solvent vapors can be ignited by brush discharge, but the probability of ignition varies with the geometry of the electrode. For example, essentially all hexane/air mixtures between the LFL and the UFL can be ignited by a spherical electrode having a diameter of 60 mm; only those hexane/air mixtures near stoichiometry can be ignited by an electrode having a diameter of 15 mm; and no hexane/air mixtures are ignited by an electrode having a diameter of 5 mm. (Heidelberg, 1967)

Those vapors having a MIE greater than hexane would be expected to be less sensitive to ignition by brush discharge; and conversely those vapors having lesser MIEs, more so. For instance, hydrogen/air mixtures are more sensitive to ignition by brush discharge than those of hexane (Heidelberg, 1967).

In cases where a brush discharge can occur at an electrode above the surface of a charged liquid, the surface potential of a hydrocarbon liquid can be used as a critical parameter for the ignition of the vapors by brush discharges. A surface potential of a negative 60 kV has been shown to be adequate (Kramer, 1979 and Johnson, 1978) but a value of "about 30 kV or more" has also been suggested (Bustin, 1983). In these cases the polarity of the liquid is very important. When the liquid carries a positive charge it induces a negative charge on the electrode and the brush discharge (if there is one) is termed a *negative brush discharge*; and conversely when the liquid is negative the discharge is a *positive brush discharge*. Positive brush discharges can be incendive to flammable vapors; negative brush discharges have not been shown to be incendive (Bustin and Dukek, 1983; and Lüttgens and Glor, 1989).

There are no substantiated cases where a brush discharge has ignited a dust. It has been postulated (Lüttgens and Glor, 1989) that the conditions for brush discharge never materialize in a dust cloud.

Incendive brush discharges are not restricted to spherical electrodes. They can occur at tips of fingers, straight or bent pipes, pipe bends, cables, casing edges, rivet heads, rims, or other conductive objects which do not have a sharp point (Heidelberg, 1967). If the grounded electrode has a sharp point, an acute edge, or a crisp corner, corona discharge rather than brush discharge will take place.

5.3.2 Brush Discharges at Surfaces

Strictly speaking, a brush discharge at a surface actually takes place in the space right above the surface; but since the spacing is so close, it is generally thought of as taking place from the surface to the electrode. The core of the brush (cf. Figure 3.2) may actually extend to the surface of the insulative material.

It is well known that discharges from the surfaces of insulative plastics and fabrics can be incendive to explosible mixtures; however, it is difficult if not impossible to reckon the energies in discharges from such insulative materials. Therefore the notion of establishing a criteria for ignition around the notion of a MIE is certainly not a straightforward endeavor. To begin with, the very notion of assigning a capacitance to a charged, insulative surface is absurd.

Nevertheless, what is needed is some sort of criteria to get a feel for the likelihood of ignition in a in-process operation. For this one can always go to an experimental duplication of an in-process condition to determine if ignition will occur or not for a particular set of circumstances. One can charge a given surface and discharge it in the atmosphere in question to see if ignition will occur or not. Such experimentation would be rather onerous even in a laboratory equipped to perform such experiments. What is needed of course is some sort of generic protocol one could use to relate such discharges to the MIE body of data.

The determination of a surface charge density is a simple task (cf. ¶ 8.4) and it is known that the maximum surface charge density in air is 27 $\mu C/m^2$ (cf. ¶ 3.2.1). In a classic set of experiments, Gibson (1965) measured the amount of charge which was transferred in discharges from plastic surfaces (i.e., brush discharges) that were incendive to some chosen flammable vapors. He then proposed the notion of an equivalent ignition energy for a given transfer of charge. In this manner one can relate, by a reasonable but unproven extrapolation, surface charge density on insulator surfaces with MIEs. Gibson's work was done with polyethylene where he was able to accumulate 11 - 23 $\mu C/m^2$ by rubbing the surface with mohair. Brush discharges were then drawn from the surface which contained up to 0.23 μC in the transfer of the charge. He then related charge transfer with the incendivity of sparks to air/acetone vapors of various stoichiometry (in the same manner as Figure 4.1). Gibson then was able to assign an equivalent discharge energy of 0.67 - 0.92 mJ for the observed brush discharges. Thus a rough (but unproven) correlation can be constructed between surface charge density and MIE.

In a later study Glor (1981) found an equivalent energy of 3.6 mJ in brush discharges from a polyethylene plate which had been rubbed with cat fur to obtain a surface charge density of the order of 10 $\mu C/m^2$.

In his conclusions Gibson states that polyethylene should not be considered as being unique and that charge densities up to 10 $\mu C/m^2$ can also be separated on Perspex®, polystyrene, polyvinyl chloride, Terylene®, nylon, and polypropylene by rubbing.

Thus, surface charge densities on insulator materials of some 40 % of the theoretical maximum can lead to discharges which can be incendive to ordinary flammable vapors.

In this regard it should be remembered that when fabrics are rubbed one upon another and remain together, such as wearing an article of clothing; surface charge densities greater than 27 $\mu C/m^2$ can be accumulated because the positive charge on one surface is in contact with the negative charge on the other. Since the surface charges of opposite sign are essentially touching, there is no external electric field to cause the onset of corona discharge in the surrounding air. When the surfaces are separated, such as removing an outer layer of clothing, then corona discharge causes the onset of sparking (they can be seen, heard, and even felt) until the surface charge density drops below the theoretical maximum of 27 $\mu C/m^2$. The remaining charge can then discharge directly in a brush discharge, be transferred to a conductive object for subsequent capacitive discharge, or be induced upon a conductive object for subsequent capacitive discharge, all of which can be incendive to ordinary flammable vapors.

5.4 Bulking Brush Discharges

Bulking brush discharge is known to be incendive to ordinary flammable gasses and vapors; however, the energy content of such discharges is not known well enough to make definitive statements in the case of dusts. Lüttgens and Glor [1989] suggest that dusts having minimum ignition energies less than 10 millijoules be considered as ignitable by bulking brush discharge while dusts with minimum ignition energies greater than 10 millijoules be considered questionable but probably non ignitable, cf. ¶ 5.4. A caveat should be added to this rule of thumb for hybrid mixtures (i.e., where both vapors and dusts are present in the surrounding atmosphere). For these cases, bulking brush discharge should be considered as being incendive.

5.5 Propagating Brush Discharges

Propagating brush discharge is an energy-rich form of a brush discharge (Glor, 1987; Lüttgens, 1985; and Heidelberg, 1967). Several joules of energy can be released in a propagating brush discharge; therefore, in industrial practice ignition of any ignitable mixture should be expected.

5.6 Corona Discharges

The concentration of molecular fragments in corona discharge is so low that only the most sensitive of vapors can be ignited; eg., carbon disulfide. Conventional wisdom has it that an optimum hydrogen/air mixture can be ignited by corona discharge, but there are no definitive references to the experimental work.

Chapter 6
Electrification

Nutshells:

[1] Streaming currents in liquids range from 1 x 10^{-10} to 1 x 10^{-4} amperes. (Cross, 1987)

[2] The range of conductivity over which streaming currents in liquids is observed is 1 x 10^{-13} to 1 x 10^{-7} S/m. (Cross, 1987)

[3] The optimum conductivity for streaming currents in liquids is about 1 x 10^{-10} S/m. (Cross, 1987; and Gavis and Wagner, 1968)

[4] After a liquid passes through a filter, its charge density increases transiently to a value 1-3 orders of magnitude higher than the steady state value. (Bustin and Dukek, 1983) (Britton and Smith, 1988)

[5] Transient (not steady state) charge densities in liquids after flowing through microfilters can be as high as 4,900 $\mu C/m^3$, (Gavis and Wagner, 1968).

[6] Streaming currents are much larger for two phase flow than for single phase flow. (Mancini, 1988)

[7] Streaming currents can also occur in insulative pipes. (Gibson and Harper, 1981)

[8] In liquids, the highest charge level observed experimentally in commercial size equipment is of the order of 1,000 $\mu C/m^3$. (Bustin and Dukek, 1983)

[9] Liquids having conductivities greater than 50 S/m (50 conductivity units) are generally considered to be nonaccumulators. (API RP 2003 and NFPA 77)

[10] *Effective* relaxation times in charged distillate oils do not exceed one minute when charge densities are high. (Bustin, 1964)

[11] When the bulk resistivity of a powder exceeds approximately 10^8 Ω·m, charging will occur in materials handling operations. (Gibson, 1981)

[12] Most organic powders have bulk resistivities greater than 10^8 Ω·m. (Gibson, 1981)

[13] For powders charged in material handling operations, 30 μC/kg is considered to be an upper limit before compacting. (Blythe and Reddish, 1979)

[14] Low conductivity powders having hydrophobic surfaces can retain charges for hours and perhaps even days. (Gibson, 1963)

[15] When two plastic sheets are separated surface charge densities of up to 2.7×10^{-5} C/m² can be attained, (Lüttgens and Glor, 1989).

[16] Significant charges can be accumulated during phase separations such as crystallization, sedimentation, evaporation, steaming, and aerosol formation. (ESCIS, 1988)

---

6 Electrification in Industrial Processes

Industrial processes are usually very dynamic where electrostatics are concerned. There are usually many competing rates of charge separation, accumulation, and relaxation. If one is to analyze the electrostatic mechanisms in such dynamic processes, one must have some quantitative idea as to the extent of charge separation, accumulation, and relaxation. In some instances there are some good quantitative models and one can analyze a process with some assurance; in other instances there are only some rules of thumb and one must do the best one can with what one has.

Many times where materials are moved about in industrial processes there are no mechanisms for separating/accumulating significant electrostatic potentials and processes are operated for years without any concern for electrostatic ignitions. However, sometimes a simple process change can lead to an unrecognized electrostatic hazard. One must therefore have some feeling for the quantities and rates of electrification when materials are moved about.

6.1 Charges in Liquids

Charges can be seperated in some liquids when they are pumped, mixed, stirred, or otherwise moved about. At the same time there is the competing mechanism of charge relaxation and when the former exceeds the latter, accumulation results. Under some conditions when liquids are splashed, sprayed, hosed, or otherwise broken up into small drops, the drops will carry a significant electrostatic charge. There are mechanisms then for charge accumulation and sometimes it may take hours for such charges to dissipate. The settling out of one liquid from another, or the sedimentation of a solid from a liquid can lead to charge separation and accumulation. The conditions under which these mechanisms can lead to electrostatic ignitions are many and varied, and the industry is continually trying to quantify the processes by which they occur.

6.1.1 Streaming Currents

When a liquid (or a powder) flows through a pipe there can be an electrostatic charge on the streaming material. When conditions are right and such charging occurs, there is a streaming current which is analogous to a current in ordinary electrical circuit.

Streaming currents in liquids are of the order of 1×10^{-5} to 1×10^{-14} amperes (Cross, 1987). As the liquid flows through a pipe, the amount of charge on the liquid reaches a steady state between the rate of charge separation at the walls and the rate of charge relaxation to the walls. The amount of charge built up on the liquid flowing through the pipe is therefore limited by the rate at which ions can diffuse to the walls and the conductivity of the liquid.

The rate at which charge flows to ground depends on the time constant of the liquid (or its conductivity); therefore a highly conductive liquid will not create a streaming current because charges can immediately flow to ground. On the other hand, a perfectly pure insulating liquid will not become charged because there are no ions to create the double layer. Thus there are two extremes where there will be little or no streaming current with an optimum somewhere in between. The range of conductivity over which charging is observed is 1×10^{-13} to 1×10^{-7} siemens per meter; and the optimum is of the order of 1×10^{-10} siemens per meter. (Cross, 1987; and Klinkenberg 1958)

An empirical relationship was developed by Schön [1957] for the streaming currents observed in flowing gasoline.

$$I_s = 3.75 \times 10^{-6} \upsilon^2 d^2 \tag{16}$$

The companion relationship for the charge density in the flowing liquid follows from the flow parameters.

$$S = 4.77 \times 10^{-6} \upsilon \tag{17}$$

A recent study (Britton and Smith, 1988) developed the more conservative (higher currents) relationships using liquids known to be electrostatically active.

$$I_s = 2.5 \times 10^{-5} \upsilon^2 d^2 \tag{18}$$
$$S = 3.18 \times 10^{-5} \upsilon \tag{19}$$

I_s ≡ Streaming current in liquid, A
S ≡ Charge density in flowing liquid, C/m^3
υ ≡ Average velocity of liquid in pipe, m/s
d ≡ Inside pipe diameter, m

It is usually of little practical importance, but for shorter lengths of pipe an additional term has been suggested.

$$(1 - \exp(-\ell/\upsilon\tau)) \tag{20}$$

ℓ ≡ Length of pipe, m
υ ≡ Average velocity of liquid in pipe, m/s
τ ≡ Time constant of liquid, s

Another streaming current equation was derived from basic principles years ago by Helmholtz (cf., Klinkenberg, 1958) which is sometimes useful, not so much for quantitative estimation but for obtaining an insight to how things vary as a function of the parameters involved.

$$I_s = A\epsilon\epsilon_0 \, \zeta \, \Delta P/\eta \, \ell \quad (21)$$

I_s ≡ Streaming current in liquid, A
A ≡ Cross-sectional area of pipe, m^2
ζ ≡ Zeta potential, V
In hydrocarbons $\zeta \leq 0.2$ V,
(Klinkenberg, 1958)
ΔP ≡ Pressure drop across pipe, Pa
η ≡ Viscosity of liquid, Pa·s
ℓ ≡ Length of pipe, m
$\epsilon\epsilon_0$ ≡ Permittivity of the liquid, F/m

It is easy to visualize the case of a insulative liquid flowing through a metal pipe. One charge will be picked up and moved along with the insulative liquid and an equal and opposite charge will flow through the conductive metal pipe to ground. In insulative pipes the mechanism is more complicated, but streaming currents are present in liquids flowing through glass, rubber, and plastic pipes which would normally be considered to be insulators; i.e., streaming currents are not limited to metal pipes.

It is known that there are certain prostatic impurities which significantly enhance streaming currents. The problem has ben studied (Leonard, 1976) where an attempt was made to identify the types of compounds responsible for unusually high electrostatic activity in hydrocarbon fuels. It was found that increasing the moisture content increased the charging tendency and in one case the charge density was increased by a factor of 23. However, it was concluded that it was not the water per se, but rather its interaction with some other constituent of the fuel, that was responsible for its prostatic effect.

These observations somewhat agree with the studies of Schön (1957) and Britton and Smith (1988) on the magnitude of streaming currents. For a conservative estimate, use the Britton and Smith relationship, otherwise use the Schön relationship; and in both cases realize that it is indeed an estimate.

6.1.2 Charge Relaxation in Liquids

For example consider a grounded metal tank containing toluene which has been charged by a streaming current as the tank was filled. After filling is complete, the rate at which the charge will relax (or find its way to ground) will be determined by the conductivity of the toluene.

Since impurities make a tremendous difference in the conductivity of dielectric liquids, the choice of the conductivity of the toluene will be a critical factor in predicting the relaxation time. Conductivities for toluene have been experimentally determined to be from 1×10^{-8} to 1×10^{-16} S/m. The dielectric constant, ϵ, of toluene is 2.38 and if the value of 10^{-8} is selected for its conductivity, then a relaxation time of 0.0021 seconds is reckoned. This value is consistent with experience.

On the other hand, if one uses the lower value of 10^{-16} a relaxation time of 61 hours is reckoned which is not consistent with experience since highly charged liquids do not retain charges for such long periods.

The exponential relation for relaxation, Equation 14, is based on the notion that a charge is transferred from one molecule to another as it makes its way to the grounded boundary. But since the ions themselves can move through the liquid, the exponential relation does not model charge migration in liquids at conductivities below about one picosiemen per meter when the liquid is highly charged. For such low conductivities at high levels of charge, the Bustin equation (Bustin, 1964) which is based on the notion of hyperbolic decay may be used.

$$S = S_0/[1 + nS_0\, t/\epsilon\epsilon_0] \qquad (22)$$

S = Charge density at time t, C/m^3
S_0 = Initial charge density, C/m^3
n = Ionic mobility, $m^2/V{\cdot}s$
(1×10^{-8} $m^2/V{\cdot}s$ for charged distillate oils)
t = Time, s
$\epsilon\epsilon_0$ = Permittivity of the liquid, $s/\Omega{\cdot}m$

This relation accounts for the experimental observation that the effective relaxation times in charged distillate oils of low viscosity do not exceed some one minute when charge densities are high. One can easily visualize that the ions in a highly charged liquid are in an electric field which is mutually repulsive to the ions which create the field. Thus, the ions near the wall of the container are forced to flow to the wall where they become neutralized. It has been suggested (Britton, 1988) that the hyperbolic decay for highly charged distillate oils is roughly equivalent to the exponential decay of an oil having a conductivity of 0.5 pS/m. Thus, combining the suggestions of Bustin and Britton regarding relaxation from distillate oils one can assume for quantitative purposes that charge decay follows the exponential form of Equation 8 for conductivities down to one picosiemen per meter. For lesser conductivities, Equation 17 should be used with a ion mobility of 10^{-8} $m^2/V \cdot s$. For qualitative purposes, effective relaxation times in charged distillate oils do not exceed one minute when charge densities are high. CAUTION: There must not be a simultaneous charging of the liquid such as sedimentation.

The ionic mobility of a liquid is a function of viscosity; therefore, for liquids of moderate viscosity the ionic mobility will be less than the suggested 1×10^{-8} $m^2/V \cdot s$. Some companies suggest that the Bustin equation be suspect for liquids having viscosities greater than 30 centistokes, for instance some diesel oils at low temperatures.

6.1.3 Liquid Conductivity

When a liquid has a conductivity greater than 10,000 pS/m it is a non-accumulator because charges will run off to ground much faster than they can be seperated by an industrial process, provided of course that there is a path to ground. Therefore, when something is known of the conductivity (or resistivity) of a liquid, Equations 14 and 17 are useful in reckoning a relaxation time and thus determining if a particular liquid will act as an accumulator. Table 6.1 contains useful conductivity data for a number of liquids. For convenience Table 6.1 has been broken up into groups of liquids in accord with their conductivity. There are many liquids which do not appear in the tables, but with care and prudence one may infer conductivities by analogy and there are some general trends.

Most chemicals have conductivities much greater than 100 pS/m, and therefore do not result in problems with static electricity in bulk liquid operations; however, there are special instances of record where electrostatic problems have arisen with liquids having conductivities as high as 10,000 pS/m. Namely, in operations where liquids were pumped through insulative conduits (eg. pipes lined with insulators and rubber or plastic hoses) and where end-of-line bag filters were used to avoid color problems from rust and scale. When a insulative conduit or an end-of-line bag filter is required while handling a liquid having a conductivity less than 10,000 pS/m, Table 6.1b, the tank atmosphere must be maintained in a non-flammable condition.

Refined hydrocarbons, the so called white oils, are considered to be accumulators of static electricity. Liquids whose chemical formula contains only hydrogen and carbon should be considered to have very low conductivities in their pure state. Impurities have the effect of increasing the conductivity of a liquid. Impurities cannot make a nonaccumulator into an accumulator.

On the other hand polar liquids are non-accumulators by their very chemical nature because they have enough conductivity to provide a path for electrostatic charges to find their way to ground.

Table 6.1a: Conductivities of Liquids

Conductive Liquids	
κ, S/m	**Liquid**
10^{-7}	1,2-dichloroethane ethyl benzoate
10^{-6}	water, methanol, ethanol n-propanol, n-butanol ethyl acetate cis-1,2-dichloroethylene
10^{-5}	acetic acid, pyridine acetonitrile, propionitrile benzonitrile, acetone butanone, cyclohexanone isobutanol
10^{-4}	isopropanol, t-butanol ethyl formate, nitrobenzene anhydrous acetic acid propionaldehyde
10^{-3}	glycol, acetaldehyde dimethyl formamide
10^{-2}	formic acid

Reference: ESCIS, 1988

It is often the case that a reliable conductivity (or resistivity) is not known. The chemical literature is not very helpful in this regard since conductivity and resistivity data are scarce at best. It is usually a futile effort to search for a conductivities or resistivities of liquids in the handbooks.

Table 6.1b: Conductivities of Liquids

Semiconductive Liquids (1,000 < κ < 10,000 pS/m			
Liquid	κ, pS/m	ε	τ, s
1° amyl acetate (4°C)*	**1250**	**4.75**	**0.034**
1° amyl acetate (23°C)*	**2160**	**4.75**	**0.020**
biphenyl (liquid @ 69-120°C)	2500-10000	n/a	n/a
bromobenzene	1200	5.40	0.040
1-bromonaphthalene	3660	4.83	0.011
iso-butyl acetate (4°C)*	**2650**	**5.3**	**0.0018**
iso-butyl acetate (23°C)*	**4320**	**5.3**	**0.011**
n-butyl acetate (4°C)*	**2170**	**5.0**	**0.020**
n-butyl acetate (23°C)*	**4700**	**5.0**	**0.0094**
butyl acrylate	3580	n/a	n/a
n-butyl propionate (24°C)*	**2300**	**n/a**	**n/a**
chlorobenzene	7000	5.621	0.0071
chloroform	<10000	4.806	>0.0043
dibutyl sebacate	1700	4.54	0.024
o-dichlorobenzene	3000	9.93	0.0029
ethylene dibromide	**<10000**	**4.78**	**>0.0042**
ethylene dichloride	**4000**	**10.36**	**0.022**
2-ethyl hexanol*	7900	n/a	n/a
methylene chloride	**4300**	**8.93**	**0.018**
pentyl acetate (23°C)*	**3400**	**n/a**	**n/a**
propionic acid (25°C)*	8500	3.44	0.0036
n-propyl acetate (4°C)*	**8460**	**8.1**	**0.0085**
sulfur (130°C)	5000	n/a	n/a
vinyltrimethoxysilane (<2% CH3OH)	5900	n/a	n/a

Notes for Tables 6.1b, 6.1c, and 6.1d: Lange's Handbook was the preferred source of tabulated conductivity data. Many data were translated from German. It was not possible to verify data in most cases. Dissipation times at κ < 2 pS/m are all based on the hyperbolic behavior exhibited by hydrocarbon fuels. Note that conductivities generally decrease with decreased temperature and with increased purity.

* indicates measured by Union Carbide Physical Properties Group - 1992/1993

Flammable liquids are given in bold type where flash-point (<100°F) was available in NFPA 325M. Flammables normally regarded as gases (such as H_2S) were excluded.

Table 6.1c: Conductivities of Liquids

Semiconductive Liquids (100 < κ < 1,000 pS/m)			
Liquid	κ, pS/m	ϵ	τ,s
armeen	470	n/a	n/a
2-ethylhexyl acrylate	610	n/a	n/a
gasoline (leaded)	**~100**	**2.3**	**~0.41**
hydrogen sulfide (@ B.P.)	1000	n/a	n/a
sulfur (115°C)	100	n/a	n/a
1,2,4-trichlorobenzene	200	4.08	0.18
trichloroethylene	800	3.42	0.037

There is one rule of thumb which is sometimes useful in this regard. There is a rough correlation between conductivity and dielectric constant and there are many liquids in the chemical literature for which a dielectric constant is cited. Therefore, if a liquid has a dielectric constant greater than ten, it can be considered a non-accumulator. This rule of thumb can be used as a first screen in those cases where dielectric values can be found. Again, some impurities in the liquid do not decrease the conductivity of the liquid.

The conductivity of a liquid decreases with a decrease in temperature and follows the following relationship:

$$\log_{10}(\kappa_1/\kappa_2) = m(T_1 - T_2) \qquad (23)$$

κ_1 ≡ Liquid conductivity at Temperature T_1, pS/m at °C
κ_2 ≡ Liquid conductivity at Temperature T_2, pS/m at °C
m ≡ Log conductivity/temperature coefficient, $°C^{-1}$

The range of 0.009 - 0.0018 for the coefficient **m** has been suggested for aviation fuels (Gardner and Moon, 1983) with the caveats that [1] it is a general trend, [2] it does not embrace different fuel types, [3] it does not accommodate to the effects of conductivity additives, and [4] it does not apply at extremes of temperature. Nevertheless, the relationship can be used as a first approximation of the expected trend.

It has been found that the charge density, which results from the pumping of aviation fuel, increases with an increase in temperature. The behavior of these fuels was quite variable, but charge densities in excess of 200 $\mu C/m^3$ were observed for some fuels when the temperature was in the

Table 6.1d: Conductivities of Liquids

Insulative Liquids (κ < 100 pS/m)			
Liquid	κ, pS/m	ϵ	τ, s
anisole (methyl phenyl ether)	10	4.33	3.8
benzene (purified)	**5×10^{-3}**	**2.3**	**~100 (dissipation)**
biphenyl (solid: less than 69°C)	0.17	n/a	not applicable
bromine (17.2°C)	13	n/a	n/a
butyl stearate	21	3.111	1.3
caprylic acid (octanoic acid)	<37	2.45	> 0.58
carbon disulfide (1°C)	**7.8×10^{-4}**	**2.6**	**~100 (dissipation)**
carbon tetrachloride	4×10^{-4}	2.238	~ 100 (dissipation)
chlorine(-70°C)	<0.01	n/a	n/a
cyclohexane	**<2**	**2.0**	**>8.8**
decalin	6	2.18	3.2
diesel oil (purified)	~ 0.1	~ 2	~100 (dissipation)
diethyl ether	**30**	**4.6**	**1.4**
1,4-dioxane (diethylene oxide)	**0.1**	**2.2**	**~100 (dissipation)**
ethyl benzene	**30**	**2.3**	**0.68**
gasoline (straight run)	**~ 0.1**	**~2**	**~100 (dissipation)**
gasoline (unleaded)	**<50 (varies)**		
heptane (purified)	**3×10^{-2}**	**2.0**	**~100 (dissipation)**
hexane (purified)	**1×10^{-5}**	**1.90**	**~100 (dissipation)**
hexamethyldisilazane	**29**	**n/a**	**n/a**
jet fuel A and A-1	0.01 - 50	2.2	0.39 - 100
jet fuel B	**0.01 - 50**	**2.2**	**0.39 - 100**
kerosene	1 - 50	2.2	0.39 - 19
naphtha	**varies**	**~ 2**	**~100 (dissipation)**
pentachlorodiphenyl	0.8	5.06	~100(dissipation)
pentachloroethane	<100	3.83	>0.3
SiH fluid (Y-10354)	2.5	n/a	n/a
stearic acid (80°C)	<40	n/a	n/a
styrene monomer	**10**	**2.43**	**2.2**
toluene	**<1**	**2.38**	**21**
trichlorosilane	**n/a**	**n/a**	**n/a**
turpentine	**22**	**n/a**	**n/a**
iso-valeric acid	<40	2.64	>0.58
xylene	**0.1**	**2.38**	**~100 (dissipation)**

Reference: Britton, 1992

range of 15°C. In other cases under the same conditions, charge densities of less than 20 $\mu C/m^3$ were observed (CRC, 1983).

6.1.4 Antistatic Additives

It is well known that some pure hydrocarbons have very low conductivities and that a small amount of impurities can increase their conductivities by one or two orders of magnitude. It is common in the petroleum industry to utilize this phenomenon by adding a *conductivity additive* to a particular product to convert it from an accumulator into a non-accumulator. When used properly this method can be quite effective in reducing the accumulation of electrostatic charges. However, these additives tend to disappear from a product as it finds its way through the distribution system such that a product may begin its journey at the refinery as a non-accumulator but change back into an accumulator before it is consumed in a vehicle.

6.1.5 Sedimentation

When a second phase, either a liquid or a solid, settles out of the liquid after it is pumped into a tank, the settling process is also a mechanism for charge separation in that the material which falls out will carry an electrostatic charge with it and leave an equal and opposite charge behind. The most common example of this process is the settling out of water from an oil but the same thing can occur when a solid settles out. Since the settling process continues after loading is complete, electrostatic charging can continue; therefore, a waiting period of at least 30 minutes is required before dipping, ullaging, gauging, or sampling operations can be performed.

When hydrocarbons are pumped into a large tank there is a streaming current associated with the flow of the liquid which results in a charged liquid within the tank. The rate of decay of this charge during the later stages of filling and after filling has been completed is slower than that which would be predicted for an exponential or hyperbolic decay; i.e., either from Equation 14 or Equation 17). The explanation for this slower decay is the sedimentation of water from the hydrocarbon.

After the hydrocarbon is pumped into a tank, the water will agglomerate into small droplets and settle out of the hydrocarbon. When they do so, they will carry with them an electrostatic charge to the bottom of the tank and leave an equal and opposite charge in the main body of the liquid. The water is conductive and the charge carried to the bottom of the tank will rapidly find its way to ground. The charge remaining on the hydrocarbon will more slowly find its way to ground and thus a significant charge can remain on the liquid.

Experience has shown that if one attempts to perform a manual operation where something is inserted into the ullage of the tank immediately after the completion of filling, one runs the risk of having a brush discharge. The standards (eg. API RP2003) state that there should be a period of 30 minutes between the filling of a tank and any manual sampling, gauging, ullaging, or instrumentation procedure.

6.2 Charges in Mists

There are a number of mechanisms whereby a charged mist can be created within an enclosure; splash filling, washing, or steaming for instance. There will be an electric field associated with the charged mist and there are operations where brush discharge can be a result. In instances where this can occur, the inerting of the enclosure atmosphere is perhaps the most desirable precaution; however, there may be cases where waiting until the mist coalesces and settles out is the preferred method.

6.2.1 Washing

During a water washing operation very high potentials are created on the mist. Under certain circumstances, discharges with sufficient energy to ignite flammable mixtures can occur from ungrounded conducting objects within, or introduced into, a tank filled with a charged mist. The processes by which ungrounded conductors give rise to ignitions in a mist are quite complex; therefore, extreme caution should be exercised when water washing operations are being carried out where flammable vapors may also be present. There should be no ungrounded conductor in the tank, and none should be introduced while the mist persists; i.e., during washing and for 5 hours thereafter.

There are occasions where liquids other than water may be required to wash a tank. In those instances the mists which are created can also become highly charged and the same precautions must be taken.

6.2.2 Splash Loading

When a liquid is splash loaded such that small droplets of a mist or an aerosol are suspended in the surrounding atmosphere, the suspended droplets can carry a significant amount of electrostatic charge; therefore, downspouts should reach to the bottom of the tank or the tank should be filled from the bottom. But even so, splashing should be minimized during the initial stages of loading by keeping the flow rate of the liquid into the tank below 1 m/s (3 feet per second). Nomogram 9.3 can be used to relate the pipeline diameter to the volumetric flow rate so that a linear flow velocity of 1 m/s per second is not exceeded. After all splashing and surface turbulence has ceased, the flow rate can be increased to the normal pumping rate of the system being used, consistent with the proper control of the operation. But, flow rates should never exceed 7 m/s (API RP2003) because of the problems with intense electrification such as "go devils"; qv.

6.2.3 Steaming

A steam jet is a horrendous separator of electrostatic charges. This is the so called "wet steam" which creates the visible white cloud. Currents of 300 μA have been observed at leaky flanges (Finke, 1989) and currents in excess of this should be expected from a steam jenny.

6.2.4 Carbon Dioxide

Carbon dioxide has a similar property when the "snow" is formed at a nozzle. Therefore, the use of carbon dioxide to inert a tank should be avoided if there is any chance that the tank to be inerted contains a flammable atmosphere at the time the carbon dioxide is injected into the tank. The "snow" of carbon dioxide can be highly charged and will create a significant electric field within the tank; therefore, there is a risk of having an incendive spark or brush discharge where there may be a flammable atmosphere.

6.2.5 Charge Decay from Mists

A typical mist in a marine tank after a washing operation is quite light and can only be seen by shining a light beam through the suspension, and it takes a rather long time for the mist to coalesce. It has been shown (Bustin, 1983) that the rate at which the space charge decays is governed by the coalescence of the mist and follows the following relationship:

$$S = \frac{S_0}{1 + S_0 kt} \tag{24}$$

S ≡ Charge density at time t, C/m³
S_0 ≡ Initial charge density, C/m³
t ≡ Time from beginning, s
k ≡ Empirical coefficient, m³/C·s
(2 - 6 times 10^4 m³/C·s)

This relationship may have its application in some specific instances; but, in general, one may be better advised to follow the *International Safety Guide for Oil Tankers and Terminals* which recommends that a period of five hours be allowed after the completion of a washing operation.

6.3 Charges in Powders

The charge separation mechanism in powders is that of frictional and impact charging while in liquids it is that of double layer charging. Therefore, there are significant differences in the behavior of solids and liquids. The movement of powders in material handling operations necessarily constitutes the rubbing of the particles against the walls of the process equipment and against themselves. It follows that the more energetic the rubbing, the more the charge separation. It also follows that the more insulative the powder (or the equipment) the more the accumulation of charge. With these trends in mind, one can see that the materials handling of low conductivity powders in industrial operations would be expected to be accompanied by electrostatic separation and accumulation problems. It is therefore important to be vigilant in accommodating the system to the potential of having electrostatic discharges.

6.3.1 Streaming Currents in Powders

There are no relationships for streaming currents in powders which correspond to those in liquids. However, the mass charge density, before compaction, in powder streams from certain operations has been ascertained, Table 6.2. These data can be used to estimate the rates of charges coming from, moving along, or going into some specific operations.

However, when powders accumulate into a pile, the charges must reorient themselves and charge compaction occurs.

Table 6.2: Typical Charge Levels on Medium Resistivity Powders Emerging from Various Powder Operations (Before Compaction).

Operation	Mass Charge Density, $\mu C/kg$	
Sieving	10^{-5} - 10^{-3}	a,b
Pouring	10^{-3} - 10^{-1}	a,b
Scroll feed transfer	1 - 10^{-2}	a,b
Grinding	1 - 10^{-1}	a,b
Micronizing	10^{-1} - 10^{2}	a,b
Sliding down an incline	10^{-1} - 10	a
Pneumatic conveying	10^{-1} - 10^{3}	b

References: [a] Mancini, 1988 [b] BS 5958

When containers are loaded or unloaded with a insulative powder the separation and accumulation of electrostatic charges on the bed of powder is to be expected.

The movement of granular material through conveyers, chutes, feeders, and other such process equipment will cause the material to become charged even though the equipment is made of metal and grounded. The simple operation of sliding an insulative material down a chute can lead to significant charging, depending upon the electrostatic characteristics of the material. When such a charged material is collected, care should be exercised to have a mechanism for the charges to dissipate. Anything that can be done to increase the rate of dissipation should be done.

A classic example of charge separation is in a pneumatic transport system. The powder is suspended in the pneumatic fluid and as the particles strike the wall of the duct, charge separation occurs. One charge remaining on the particle and the other on the duct. The charge tends to remain on the particles as they move through the system because charge cannot relax through the transport medium. When air or nitrogen is used, there is no effective way for the charges on the particles to find their way back to ground and significant streaming currents result. There are no good estimation techniques for approximating a streaming current in pneumatic transport operations or in any two phase flow. In these instances, experimental determinations must be made to quantify ignition hazards. In practice, a streaming current of 100 μA is considered to be a maximum (BS 5958).

In pneumatic transport there is a lot of energy put into the movement of granular materials and powders. It is therefore to be expected that very high charging and charge accumulations will result. Associated equipment should be conductive and properly grounded. There have been instances where plastic pipe has been substituted for metal pipe. This may be possible in some applications; but in applications where the powders are electrostatically active, propagating discharge may be a result. (cf. Case Histories)

A fix which has been used on occasion is that of maintaining a high humidity in the pneumatic transport fluid. In some instances this has made a difference of three orders of magnitude in the amount of electrostatic charging.

In most pneumatic transport systems there are bag houses (filters, filter houses, collectors) installed to collect the fine dust resulting from the manufacture or the movement of the product. In such cases there is an energetic electrostatic separator (the pneumatic transport system) and a location where electrostatic charges can accumulate in the presence of a ignitable dust (the bag house). A common failing in bag houses is that of keeping the cages properly grounded so that they do not act as accumulators of dangerous electrostatic potentials. One method which has been successful is that of sewing two braided ground straps over the top of the cuff of the bag. See Figure 10.11.

6.3.2 Charge Compaction in Powder Bulking

When powders or granules of product emerge from process equipment they will carry electrostatic charges on individual particles in the stream. Consider a stream of charged particles which are initially well separated from each other and have maximum surface charges, Table 6.2. The upper limit of charge density will be determined by the dielectric strength of the air surrounding each granule. This charge density can prevail in the stream as it moves through the process and the granules remain separated.

At the end of the stream where the granules are being collected into a pile, the forces of gravity are greater than the electrostatic forces of repulsion and the granules are compacted. As compaction occurs the charges reorient themselves such that the charges on the surfaces of each of the small granules migrate to the surface of the agglomerates and as the agglomerates are compacted the charge migrates to the surface of the pile. The effective surface area of the whole pile is much less than the sum of the individual surface areas. (Take the ideal example of 10^{12} cubes 100 μm on an edge being stacked into a cube 1 m on an edge; the surface area decreases by a factor of 10^4.) This decrease in surface area leads to an increase in the surface charge density of the pile causing the dielectric breakdown strength of air to be exceeded. Ionization of the air occurs at the surface of the pile where an immense amount of charge is released. The ions in the air will then neutralize enough of the surface charges to cause the field to fall below the ionization threshold. This process is cyclical and as the filling process continues the ionization from the bulking process is repeated.

Because of the ionization which accompanies the bulking process, the charge density in a compacted pile of powder is less than that which would be inferred from Table 6.2.

6.3.3 Charge Relaxation in Powders

Relaxation times in powders differ significantly. Low conductivity powders having hydrophobic surfaces can retain charges for hours and perhaps even days. Since the particles themselves cannot migrate toward the boundary as can the ions in a liquid, there is no analogy to the hyperbolic relationship in powders. Antistatic additives can be used in powders in some instances, but there are no generalizations which govern their use.

The important factors governing the relaxation of a charge are the electrical conductivity of the charged material and the resistance through any material interposed between the charge and ground.

6.4 Surface Charges

6.4.1 Triboelectric Charging

Different materials have different propensities for giving up or accepting electrons. For instance if a silk cloth is rubbed on Teflon™, the silk will become positively charged and the Teflon will be negative; but if the silk cloth is rubbed on mica, the inverse will occur. Mica gives up electrons to silk while silk gives up electrons to Teflon; or looking at it the other way around, Teflon captures electrons from silk, but silk captures electrons from mica. Different insulators have different affinities for carrying a charge and can be ranked according to their affinity for positive or negative charges. Such a ranking is the *triboelectric series*, Table 6.3, where materials are ranked according to the sign of charge they will carry. The further apart materials are in the triboelectric series, the more charge separation will occur for a given set of conditions.

Table 6.3. The Triboelectric Series

Material
+
Mica
Wool
PVA
Polymethyl Methacrylate (Plexiglas[tm])
Nylon 66
Silk
Viscose
Cellulose Acetate
Cotton
Polystyrene
Polyvinyl Butyrate
Dacron[tm]
Polyacrylonitrile (Orlon[tm])
Polyethylene
Polytetrafloroethylene (Teflon[tm])
Polyimide

Reference: Cross, 1987

One may therefore be tempted to conclude that charging would not occur if two identical materials are rubbed together. But such is not the case because the surfaces will invariably be contaminated to some degree and surface asperities make the surfaces different. Therefore, some amount of charging is to be expected between surfaces of the same material. A classic example of this is the unrolling of a sheet of plastic film.

Even with pure, clean crystals each of the crystallographic surfaces have different electrical properties and triboelectric charging can occur.

6.4.2 Humidity

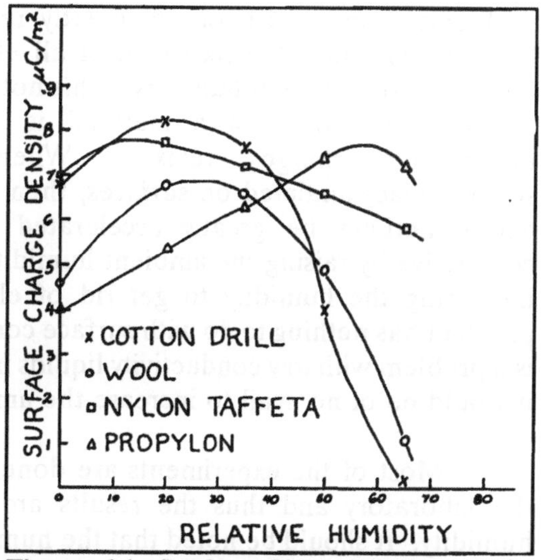

Figure 6.1: Surface Charge Density as a Function of Humidity (Sereda and Feldman, 1964)

At zero humidity a clean, insulative surface can be charged by rubbing with an appropriate material, and for a given amount of rubbing there will be a particular charge density. So at zero humidity there are intrinsic sites for ions to reside. As the humidity begins to increase, water molecules will be adsorbed on the surface and act as additional sites. Thus, for a given amount of rubbing there will be an increase in charge density for an increase in humidity at low humidities. This process prevails until a monomolecular layer of water is formed on the surface. At this point the surface conductivity increases with a further increase of humidity. This increase in conductivity provides more paths for charges to migrate and results in a lesser amount of surface charge for a given amount of rubbing.

A classic experimental demonstration of this mechanism was shown by Sereda and Feldman (1964) for a few chosen fabrics. They did not report the data in terms of surface charge density but did give the information necessary to reckon them, Figure 6.1. They demonstrated that the maxima of charge density occurred when a monomolecular layer of water was completed on the surface of the fabric. These maxima corresponded to a charge density of $\sim 8 \times 10^{-6}$ C/m^2 (some 30% of the theoretical maximum) and were the same for all fabrics tested. Note that similar charge densities can be impressed on hydrophobic fabrics at relative humidities of 65% as can be impressed on other common fabrics at relative humidities of 20-30%!

As the above discussions show, an increase in humidity can be used to decrease the amount of electrostatics accumulated in a system - *provided it is surface conductivity related*. A very common misconception exists about the effects of humidity upon electrostatics. It is thought that electrostatic problems cannot exist, or are greatly reduced, during periods of high humidity and that the fix for most all electrostatic problems is that of increasing the ambient humidity! This notion has credence only where the surface conductivity of a material is the mechanism by which the electrostatic charges relax. When electrostatic charges are separated/accumulated on surfaces, then the relaxation of those charges can sometimes be greatly accelerated by making the surface more conductive by raising the ambient humidity. Mistakes are often made by increasing the humidity to get rid of electrostatic problems when the problem has nothing to do with surface conductivity. For instance, if there is a problem with low conductivity liquids in pumping operations, obviously it would be of no avail to increase the ambient humidity.

Most of the experiments are done at the ambient temperature of the laboratory and thus the results are reported in terms of relative humidity. It should be noted that the humidity effects are that of absolute humidity rather than relative humidity (eg. Katrak, 1995).

6.4.3 Conductive Cloth and Plastics

Cloth can be made conductive by the addition of conductive fibers into the makeup of the cloth. Many variations have been used; eg., conductive threads have been woven into both the warp and weft of the cloth, conductive fibers have been used in the making of felts, metal wires have been woven into the fabric, etc. The efficacy of these cloths in some applications is in serious question.

Plastics can be made conductive by the addition of carbon into the formulation. The result is a black plastic (not all black plastics are conductive) and when implements are made of such plastics, they must be grounded.

6.3.4 Neutralizers

One way to eliminate electrostatic charges from a surface is to ionize the air directly above the surface. There are many ionizers on the market which can be used in various applications; however, one must get the ionized air to the place where the charges reside, something that is not always possible to do in a chemical process.

6.5 Intense Electrification

The discussions thus far about electrostatic phenomena have to do with modest amounts of charge separation and charge dissipation; however, there are instances where charges are separated much faster than they can be dissipated. This can either be from a very high rate of separation where large amounts of energy are being put into the system in a very short time or from a very low rate of dissipation where things are well insulated from ground. In these cases potentials and gradients rise until the breakdown of the surrounding medium (usually air) is exceeded and ionization and some sort of discharge occurs. It is instructive to examine two examples.

[1] When an insulative liquid is being pumped into a tank at a very high rate; eg., linear flow ⩾ 7 m/s (API RP2003), discharges can be observed on the surface of the liquid. As flow is increased, a corona type discharge is first noted. As flow is increased further, discharges across the surface of the liquid are observed; and at high flows, discharges two to four feet long can be observed. In these cases a tremendous amount of electrostatic charge is created by streaming currents. When this charge is put into the body of the liquid in the tank, the mutual repulsion between the unit charges forces migration of the ions to the boundaries of the liquid. Those which go to a metal wall can easily find their way to ground, but those which go to the surface of the liquid have no place to go and create a high surface charge density (Klinkenberg, 1958). The field at the surface soon exceeds the dielectric breakdown strength of air and discharge across the surface occurs. These high energy events are called "go devils". A similar phenomenon can be observed in insulative (eg. glass lined) vessels during vigorous stirring.

[2] When two dissimilar insulative materials are in intimate contact, there may be a very high density of charge on each of the surfaces - positive on one surface and negative on the other. But since they are in intimate contact, the opposite charges effectively neutralize each other so that there is no external field and the density of the pairs of charge may far exceed the maximum charge density discussed in ¶ 3.2.1. This condition can exist indefinitely as long as the surfaces remain in contact. If such is the case and the surfaces are rapidly separated, the surrounding air will ionize. Examples are when powders are rapidly dumped from plastic containers, when one fabric rubs across another, when clothing is removed, when two plastic sheets are separated, etc. Depending upon the degree of electrification, small discharges can be heard, seen, and even felt; eg., the crackling heard and felt when sliding across a car seat during

periods of low humidity. After the ionization process and corona discharge is completed surface charge densities of up to 2.7×10^{-5} C/m² can be attained, (Lüttgens and Glor, 1989). Residual charge densities of 1×10^{-5} C/m² are not uncommon after vigorous rubbing or separation of insulative sheets and films.

In industrial situations where procedures and designs are in place to take care of ordinary electrostatic problems, the standards give some additional caveats such as upper limits of flow rates. These upper limits (eg. 7 meters per second in API RP2003) are imposed as additional operating criteria because if all the other guidelines are strictly adhered to, there will still be the problems of intense electrification; eg., brush discharges, "go devils", etc.

6.5 Phase Separation Charges

Charge separation should be expected during phase separation operations such as sedimentation, crystallization, aerosol formation, and even evaporation (ESCIS, 1988). There are no good rules of thumb for estimating the amount of charge separation to be expected in these types of operations. However, one should consider these operations to be charge separators to some degree.

Chapter 7

Design and Operating Criteria

Nutshells:

[1] Thou shalt have no ungrounded conductors in any area where there is a possibility of having an ignitable mixture.

[2] A resistance of a conductor to ground of less than one megohm is adequate for an electrostatic ground. (eg. NFPA 77)

[3] In new or modified equipment where a metallic path to ground is to be relied upon, a resistance of less than ten ohms should be required. (Walmsley 1992)

---

7 Design and Operating Criteria

In an ideal world, all the electrostatic problems could be made to go away by keeping everything conductive and keeping all equipment grounded. Unfortunately, things cannot be that simple because there are many insulative process materials and insulative pieces of process equipment. Therefore, there will necessarily be places in industrial processes where there will be charge accumulations. Thus the engineering approach is to prevent charge buildup by providing a path for them to dissipate rapidly. {If you don't want a lake, don't dam the creek.} Failing that, decrease the further generation of charge. {If you are in a hole, quit digging.} Failing that and charge has accumulated, cease all other operations until the charge has dissipated. {Let sleeping dogs lie.}

7.1 Grounding and Bonding

The grounding of a conductive object is achieved by establishing an electrically continuous, low resistance path between the conductive object and the general mass of the earth. (In this context perhaps the British term of "earthing" is more descriptive.) Objects may be inherently grounded through intimate contact with a series of other objects which are in contact with the general mass of the earth, or a deliberate electrical connection can be made between an object and the earth. The baseline potential is the potential of the earth which is taken to be zero potential; thus, by definition, the potential of the earth is always zero. Therefore, all conductive objects which have the required electrical connection to the earth will be at zero potential and there cannot be sparking between them. In a closed system it follows that if everything is conductive and grounded there can be no electrostatic initiation of an ignitable mixture.

The question now arises as to what constitutes an adequate ground. It is here that some confusion results because investigators may sometimes want to translate what they know in some other area to the area of electrostatics. This sometimes leads to the erroneous notion that a very low resistance is required for an electrostatic ground when the fact is that a resistance to ground of less than $10^6\ \Omega$ is always adequate and in most cases less than $10^8\ \Omega$ is adequate for an electrostatic ground.

The $10^6\ \Omega$ criteria comes from experience. In a normal industrial operation a potential of at least 300 V is considered to be necessary for an incendive discharge, but for explosives a potential of 100 V is considered to be necessary (BS 5958). Also, it is known that charging currents in industrial operations do not exceed 100 μA (BS 5958). Therefore, from Ohm's Law, Equation 1, and the equivalent circuit of Figure 2.5, a resistance of $10^6\ \Omega$ will always be adequate to keep potentials below 100 V. Furthermore, in most industrial situations charging currents cannot exceed 1 μA and in these cases a resistance to ground of $10^8\ \Omega$ is adequate.

Bonding is where a suitable electrically continuous path is established between conducting objects (cf. Figure 1.13). In the case of bonding, the objects are maintained at the same potential and there can be no spark between them. *BUT*, they may not be at ground potential and there could be scenarios where there could be a spark between the objects and ground! In such operations redundant grounding and bonding is usually recommended by operating companies.

When grounding or bonding is to be provided by an entirely metallic path through process equipment, a resistance of no more than 10 Ω should be required at the time of installation or maintenance. This is because a resistance of more than 10 Ω indicates that the intended metallic path is not properly established because of such things as corrosion or loose connections. Therefore, the resistance may be less than one million ohms at the time of the measurement but it cannot be relied upon to remain so. This 10 Ω criterion can be waived if it is known why the resistance is higher and it is certain that the resistance is stable and will not subsequently rise above one million ohms. The notion here comes from experience (Walmsley, 1992) and is that if the resistance is less than 10 Ω at installation, it will never exceed 10^6 Ω over the life of the equipment.

7.1.1 Insulation from Ground

If a material has a comparatively high conductivity, charges can dissipate rather quickly such that no charges are accumulated; provided there is a conductive path to ground. But, if the material is isolated from ground by means of an insulator, the charge can be accumulated on the conductive material. The relaxation time for the dissipation of the charge will then be dependent upon the conductivity of the insulator.

The important factors governing the relaxation of charges from the separated materials are therefore the electrical conductivities of the charged materials and of any additional materials interposed between them and ground after their separation.

7.1.2 Spark Promoters

Any unbonded conducting body can accumulate an electrostatic charge, either by direct contact or by induction, and must be considered a *spark promoter*. Likewise, any grounded conducting object protruding into an electric field will distort the field in such a manner as to promote the onset of a one electrode discharge and must therefore also be considered a *spark promoter.*

Care should be exercised to avoid spark promoters within a tank compartment wherever possible. Tanks should be inspected and any unbonded objects removed before loading operations are begun. Where flammable atmospheres are to be expected conductive devices, such as gauge tapes, sample thieves, or thermometers should not be lowered into or suspended in a compartment, either during filling or immediately afterward.

Walmsley 1992 reports that...."*flexible conductive hoses that had metal reinforcing wires. Tests showed that hazardous potentials could accumulate on these wires if they were insulated from earth. The hoses are designed so that the wires should be earthed via the hose couplings but a survey showed that this was not always the case. Another company using similar hoses subsequently reported the occurrence of large sparks on the outside of a hose with a reinforcing wire that had accidently become insulated from earth...."*

7.2 In-process Relaxation Times

7.2.1 Quiescent Relaxations

Many times it is the result of an operation that charges have accumulated on insulative process materials, whether the materials be atmospheres, liquids, or solids. In these cases it is necessary to wait for the charges to dissipate before any other operations are carried out. There are experience based recommendations in the standards regarding the time one should wait before continuing with a procedure.

API RP2003, NFPA 77, and ISGOTT state that a period of 30 minutes should be provided after the completion of the filling of a tank before performing any manual operation; such as gauging, ullaging, dipping, sampling, or instrumenting. This period of 30 minutes is recommended because of the possibility of having brush an incendive discharge when an object is inserted into the tank where there may be an electric field. Brush discharges occur at grounded electrodes; therefore, keeping the item grounded is not a fix. The 30 minute period is considered by some as being a bit conservative; but there have been cases where the sedimentation process is long enough to warrant such a period.

In large marine tanks where a charged mist has resulted from a washing (or similar) operation, the time required for the charge to dissipate by coalescence is much longer. ISGOTT specifies that a period of 5 hours be provided after the completion of an operation where a mist was created in a tank and the insertion of an object into the tank.

7.2.2 Relaxation Downstream of Filters

Where a insulative liquid is pumped through a filter, there can be horrendous generation of charges which will remain on the liquid downstream of the filter. It is not desirable to pump such a liquid into a tank while the liquid is still carrying such a high charge. Therefore, the filter should be placed upstream of the tank such that the charges have time to relax to the walls of the pipe before they enter the tank. API RP2003, NFPA 77, NPCA 803 specify a residence time in the downstream pipe before entering the tank to be 30 seconds. This criterion is achieved by adjusting the pipe diameter and the flow rate, cf. Nomogram 9.3.

BS 5958 specifies a residence time of 3τ for liquids with conductivities down to 2 pS/m, and for less conductive liquids, 100 s. This criterion is reached with the aid of Equation 10 when the dielectric constant and the conductivity of the liquid is known. For a hydrocarbon having a dielectric constant of approximately 2.5 and a conductivity of 2 pS/m, a residence time of 30 s is reckoned. This squares with the other referenced standards but goes a step further for the liquids having conductivities below 2 pS/m, for which 100 s should be allowed.

It should be noted that these specifications have their genesis in operations with hydrocarbons. One should therefore take care in relying upon them for those few insulative chemicals having larger dielectric constants. It has been suggested (Britton, 1992) that if 30 s applies to hydrocarbons then some 100 s should be applied to chemicals under similar situations.

7.3 Simultaneous Operations

It follows from the above that there should be no simultaneous operations within tanks where there is any possibility of having flammable vapors. That is, one should not insert any conductive object into a tank while there is an operation going on within the tank. One should not do any dipping, sampling, gauging, ullaging, instrumenting or otherwise insert a conductive object into the outage space of a tank while there is a filling, pumping, washing, steaming, stirring, or other operation going on within the tank which could result in the charging of the liquid.

If there is the dynamic situation where a charge can be generated and accumulated on the liquid, one does not want to stick something into the tank where brush discharge can take place.

A corollary to this is that one does not want to have a permanent conductive fixture in the tank which can act as a site for brush discharge as an electric field is being built up by an operation within the tank. The classic example of this is a permanent metal gauging rod sticking down into the tank which acts as an electrode to the rising level of charged liquid as the tank is being filled.

Another corollary to this is having a loose conductive object within the tank which can act as an accumulator; such as a floating sample can.

7.4 Sounding Pipes

Sometimes it is an operational necessity to perform simultaneous operations, the taking of first end samples for instance. This can be done safely if the tank is equipped with a sounding pipe - sometimes termed a stilling well. This is a pipe with many small holes in its wall which extends from the top of the tank to the bottom of the tank and makes positive electrical contact with both the top and the bottom of the tank; eg., welded. Because of its small cross section, the electric field within such a pipe will always be low even if the surrounding liquid carries a significant charge; therefore, an object inserted into it will be shielded from the rest of the tank. When sounding pipes are used the operations of dipping, sampling, gauging, ullaging and, instrumenting can be carried out at any time during an operation involving liquids of low conductivity.

Chapter 8

Measurements

Nutshells:

[1] In general, ordinary multimeters can not be used to make electrostatic measurements.

[2] Measurements are quite variable among individuals, but fields of approximately 1,000,000 V/m are detectable by the hairs on the forearm. (10,000 volts/cm.)

---

8 Measurements

The measurement of electrostatic parameters is much different than that of measuring ordinary electrical parameters. For instance in electrostatics the quantities are kilovolts and microamperes while in electrical circuits they are volts and amperes. Different equipment is required to measure these quantities.

8.1 Multimeters

When a person who is unfamiliar with the science of electrostatics is faced with the necessity of making some in-process measurements, the natural tendency is to use familiar equipment which has been designed to measure current electricity. One must remember that the quantities of voltage, amperage, and resistance differ by orders of magnitude between electrostatics and current electricity. (An ordinary light bulb may draw one ampere while a streaming current of one nanoampere may be significant - a difference of nine orders of magnitude.)

The standard hardware store dc multimeters usually have a maximum capability of around 30 megohms at potentials of 1.5 volts; therefore, they are quite inadequate for measuring high resistances. They can be used however to check grounding and bonding connections in some instances. If a pocket multimeter shows continuity, then one is assured that a ground connection exists as far as electrostatics is concerned; but if continuity is not indicated, the result of >30 megohm will provide no useful information about the relaxation time of any charge which may be present.

Since an ordinary multimeter draws a significant current to maintain the deflection of the needle it cannot be used to detect the presence of an electric field. If there were a surface charge on a dielectric creating an electric field and a multimeter were touched to the surface, the amount of current drawn from the surface would not register on the instrument.

When measuring resistance to ground to assess relaxation times, instruments capable of measuring 10^{10} ohms to 10^{14} ohms are required at potentials of 100 volts or greater.

Insulation testers and "meggers" are used to check the insulation in electric motors and are sometimes useful in measuring resistances for electrostatic purposes. They impress a potential of 500 V across a test item and can indicate resistances at $> 10^9\ \Omega$. On some instruments a movement of the needle which is just perceptible indicates a resistance of the order of $10^{10}\ \Omega$. These instruments provide a middleground between an ordinary multimeter and an expensive teraohmmeter.

8.2 Electrometers

An electrometer is a highly refined dc multimeter, and can be used for virtually any measurement task performed by the more conventional equipment. The voltmeter function has a very high input resistance, up to 10^{16} ohms, and the input offset current may be as low as 5×10^{-17} amperes; therefore, voltages can be measured with a very small amount of circuit loading. The ammeter function can measure currents as low as 10^{-15} amperes. The ohmmeter function can measure resistances up to 10^{14} ohms. When used in the coulombmeter mode, currents are integrated and charges down to 8×10^{-16} coulombs can be detected. In making measurements, electrometers do draw some small amount of charge; they therefore alter the quantity being measured to some degree.

8.3 Electrostatic Voltmeters

A contact electrostatic voltmeter draws a small amount of charge to get a needle deflection but draws no charge to maintain a needle deflection. An ordinary multimeter on the other hand, draws a current to maintain a needle deflection and therefore cannot be used to measure an electrostatic potential. In using volt meters of this sort the probe is actually contacted to the conductive surface to be measured. When contacted to charged insulative surfaces, no potential is indicated on the meter. The pseudo potential of a insulative surface can be inferred by the use of a "noncontact voltmeter" or a fieldmeter.

8.4 Fieldmeters

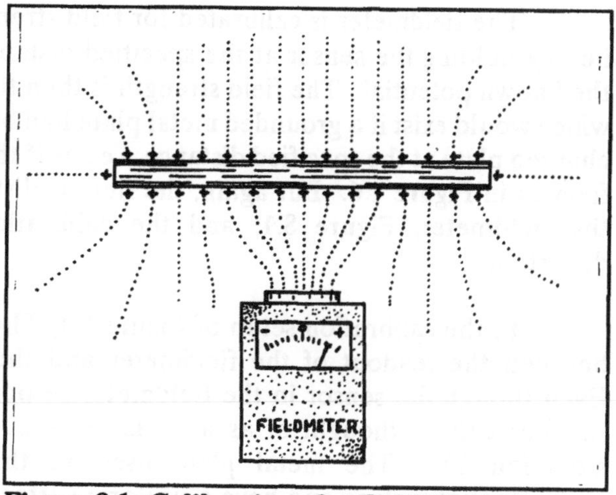

Figure 8.1: Calibration of a Field Test Meter

A fieldmeter (also known as field test meter, field mill, electrostatic locator, or static sensor) is quite useful in detecting the existence of an electrostatic field. Some of the units are inexpensive and are worth considering as an everyday tool. The fieldmeter responds to the strength of the electrostatic field into which it has been placed. It can be looked upon as an instrument which places a test charge into an electric field and measures the force on that charge. It would seem then that a measurement would be easy and straightforward, but such is not the case. The measurement seems to always be bewildering to the neophyte and sometimes misused by the seasoned. This may be because of the way the instruments are calibrated.

The fieldmeters are calibrated in two ways: in units of potential (volts) or in units of field strength (volts per distance). When they are calibrated in units of potential, the readout is volts or kilovolts. When they are calibrated in units of field strength, the readout can be in volts per inch, in volts per centimeter, in kilovolts per meter, or combinations thereof. If the unit is calibrated for potential, the instructions for the unit will specify that the sensor be held at a specified distance from the surface of the object being measured, a few inches or centimeters.

The fieldmeter is calibrated for potential by placing a potential on a metal plate while holding the sensor at the specified distance; eg., 2 inches. The meter scale or the digital readout is calibrated to conform to the known potential on the plate. It must be remembered that the field meter distorts the electric field into which it is placed, Figure 8.1; and the calibration allows then for the distortion. Therefore, the calibration plate must be large as compared to the field meter, say 16" x 16", so that the field distortion can be compensated for.

The fieldmeter is calibrated for field strength in the same manner; i.e., by holding the sensor at the specified distance from the plate having the known potential. The field strength is then that value of field strength which would exist if a grounded metal plate had been placed parallel to the charged plate at the specified distance, i.e., as if there were a homogeneous field as in Figure 1.6. But again, the field is distorted by the presence of the fieldmeter, Figure 8.1; and the calibration compensates for the distortion.

In the calibration setup of Figure 8.1, There is a direct correlation between the readout of the fieldmeter and the surface charge density. Even though the sensor in the fieldmeter responds to field strength, the readout can be thought of as a measure of surface charge density, cf., Equation 11. The metal plate used in the calibration will be a equipotential surface and have a uniform surface charge density.

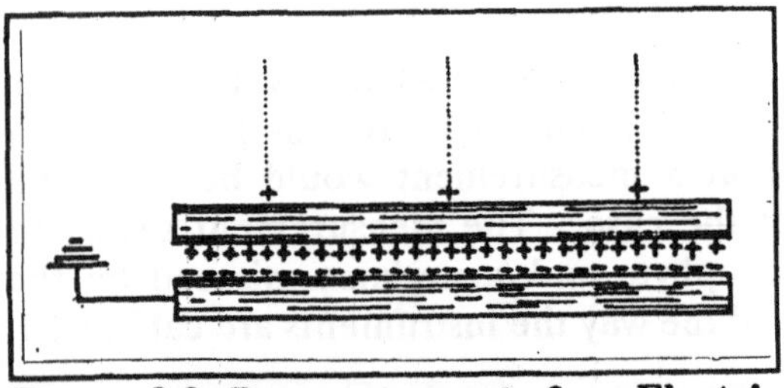

Figure 8.2: Rearrangement of an Electric Field around a Calibration Plate

A further complication must now be considered. The proximity of other grounded objects can alter the measurement considerably. Even when the charge on the conductive object being measured remains constant, the potential or field can vary if other conductive objects are in the vicinity. For example, the metal sheet used for the calibration in Figure 8.1 is in "free" space. If the same sheet with the same total charge were to be backed by a grounded metal plate, the charge would reorient itself such that the surface charge density on the "front" of the plate would be much less, Figure 8.2. A field meter would then indicate a much smaller potential or field because of the smaller surface charge density on the "front" of the plate.

Notice that the instrument [1] senses the strength of the electric field at the specified distance, [2] accounts for the specified spacing, [3] accounts for the distortion of the field, and [4] converts the measurement to potential and/or field strength. It can be seen then that an accurate measurement can only be made when the following conditions prevail:

- A flat surface.
- A large surface; i.e., large compared to the fieldmeter.
- A uniform charge density; i.e., a metal surface.
- A free standing surface; i.e. not near other grounded objects.

Field meters are handy qualitative instruments to determine where charges have accumulated and to get an idea as to how much charge has accumulated. Quantitative approximations can be useful as long as one takes reasonable care to accommodate for the deviations from the ideal case of the calibration. For instance, when measuring an electrostatic charge bound to a insulative surface, the notion that the surface is at the indicated potential is erroneous. It is intellectually dishonest to even consider that a plastic surface have a potential. However, if one can assume a reasonably uniform surface charge density, a measure of the field strength can lead to a reasonable estimate of the surface charge density through the use of Equation 11.

Another example would be the use of a fieldmeter to assess a space charge, such as a charged mist within a tank, Figure 1.11. A fieldmeter could be used at a tank opening to measure the field strength at the wall, Figure 1.12. To do this one should hold the fieldmeter at the center of an opening at the tank wall. Not down in the tank or away from it. In this manner one can get a reading which should be close to the field strength were the field meter not there and the opening closed. But, if one has a fieldmeter of the sort that is calibrated only for potential, one would be at a loss to make sense of the reading other than to say that there is an electric field present. A measure of potential at the tank wall means nothing since the potential at the wall is zero, Figure 1.12. Thus, a potential type fieldmeter would yield a misleading answer.

Another example where caution would be advised is in the assessment of the amount of charge on granules moving on a conveyor belt, Figure 8.3. Fieldmeter A would give a much lower reading than Fieldmeter B because of the proximity of the grounded roller at position A. If one did not account for the effect of the roller on the electric field,

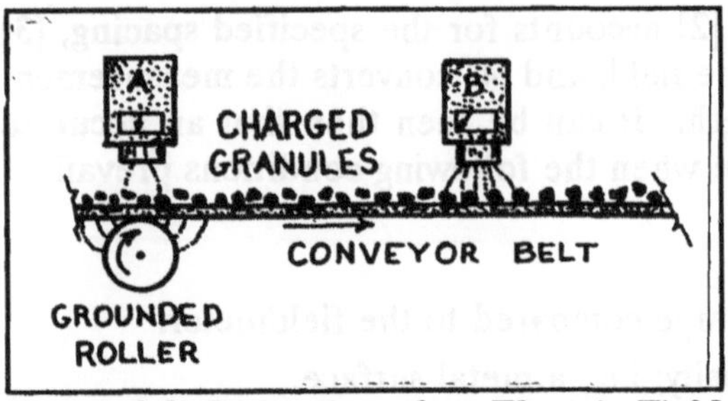

Figure 8.3: Distortion of an Electric Field by Grounded Process Equipment

one might think that the increase in field strength was due to a charge being put on the belt by the contact and separation of the belt from the roller. Thus, if one used the reading from Fieldmeter A, it would lead to an erroneous under assessment of the amount of charge on the granules.

A very crude but sometimes effective way of sensing the presence of an electrostatic field is by feeling ones hair "stand on end". The measure is quite variable among individuals, but fields of approximately 1,000,000 V/m are detectable by the hairs on the forearm. Note that this is an electric field strength of approximately one third of the electric breakdown strength of air at 3,000,000 V/m. Therefore, if an electric field can be sensed by "feeling it", it is evidence of intense electrification somewhere in the process.

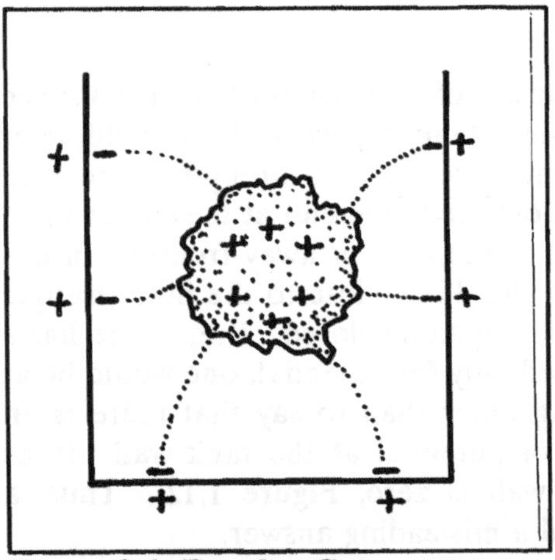
Figure 8.4: Faraday Cage

8.5 Faraday Cage

A Faraday cage is an isolated conductive enclosure into which an object can be placed. If the object carries an electrostatic charge, the charge will be induced (or transferred) to the outer skin of the conductive enclosure, Figure 8.4. The outer skin then becomes an equipotential surface and its voltage can be reckoned by using a fieldmeter. The amount of charge on the object is then reckoned from the measured voltage and the capacitance of the enclosure; Equation 8, $Q = CV$.

In the original experiments performed by Michael Faraday, he used his laboratory ice pail, and to this day similar experiments are termed "ice pail" or "cage" experiments. Following Faraday, one can perform crude experiments to determine the amount of charge on a material. Take an appropriate size bucket and place it on a sheet of clean Teflon^{tm}. Place a field test meter at its specified distance for measuring voltage and zero the meter with the bucket grounded, Figure 8.4. Remove the ground wire so the bucket is isolated and place the charged material in the bucket. Read and record the voltage on the meter. Estimate the capacitance of the bucket (cf. ¶ 9.1.1) and calculate the charge on the object, Q = CV. The volume charge density or mass charge density can then follow from a measure of the volume or the mass of the material. Such crude experiments will sometimes incur the disbelief of the nit pickers but they can yield useful approximations.

A similar experiment can be used to estimate the relaxation time of the material, (cf. ¶ 6.3.3). Instead of "looking" at the bucket, "look" down on the material and ground the bucket. Monitor the rate at which the charge dissipates from the material. The relaxation time, τ, can then be approximated from Equation 8.

Figure 8.5: Crude Faraday Cage Experiment

$$Q = Q_0 \exp(t/\tau)$$

This is done by reckoning an initial charge, Q_0, and waiting for some charge dissipation to occur. At a time, t, the remaining charge, Q, is reckoned and a relaxation time, τ, follows from the relation. A semilog plot of charge vs. time is also a convenient way to determine τ.

8.6 Radios

Another device which is useful in surveying some processes is a small radio. When electrostatic discharges occur, an electromagnetic disturbance is heard on the radio as static. The radio is tuned to a frequency in the AM band where there is no other signal and taken into an area where electrostatic discharges are suspected. By "listening" to the radio, static is evidence of electrostatic discharge. With a modest amount of experience one can use this technique to monitor some processes. Fluorescent lights and commutators are a nuisance.

Chapter 9

Quantification of Electrostatic Scenarios

Nutshells:

[1] The capacitance of a free standing spherical conductor in picofarads is approximately equal to its radius in centimeters. (Haase, 1977) For a spherical conductor near a ground plane its capacitance in picofarads is approximately equal to its diameter in centimeters.

[2] The surface resistivity of an insulative material may decrease by a factor of 1 x 10^6 in going from 30% to 90% relative humidity. (Gibson and Lloyd, 1963)

[3] Little or no electrostatic effects occur on surfaces at relative humidities greater than 75-80% except on surfaces which are markedly hydrophobic. (Gibson and Lloyd 1963)

[4] In solids, experience has shown that high surface charge densities of 10^{-7} to 10^{-5} C/m² are reached on the particles, but at charge-to-mass ratios above 3 x 10^{-6} C/kg particles no longer compact because of electric field effects, (Lüttgens and Glor, 1989).

9.1 Approximations

Many times the likelihood of an ignition by a spark can be estimated from the process parameters. Usually the first approach is that of assuming some sort of worst case condition and make a "back of the envelope" calculation to approximate the order of magnitude of the supposed spark. One can then look at the reasonableness of the assumptions used, and assess what additional steps should be taken to vindicate the postulated electrostatic scenario. In some cases it may be obvious, in other cases additional information may be necessary or testing may be required. The first shot is the ballpark estimation.

In general, the fundamental property of interest for assessing electrostatic spark discharge hazard is the potential energy of the spark. If the discharge is capacitive in nature, energies may be estimated from Equations 12, 13, 14, and 15.

$$W = CV^2/2 = QV/2 = Q^2/2C \quad \text{and} \qquad Q = CV$$

W ≡ Energy, J
C ≡ Capacitance, F
Q ≡ Charge, C
V ≡ Potential, V

Strictly speaking these relations apply to the energy stored in the capacitor and not the energy in the spark discharge, since some of the energy is lost in the discharge circuit. Nevertheless, if two out of the three quantities of charge, capacitance, and potential can be estimated, then an estimation of spark discharge energy follows from these relations. This estimate can then be compared to the MIE of the explosible mixture for an assessment of the potential hazard. It should be noted however that, in general, the in-process discharge energy for spark discharge from ungrounded process equipment is usually much greater than that required to ignite flammable vapors; i.e., if the postulated scenario is that of an ungrounded conductor, the discharge energy is most likely many times the MIE of the material.

When doing the first cut approximation, one does not need precise quantities to plug into the equations. Some quantities may only be known to within one or two orders of magnitude, or perhaps three, while others may be reasonably estimated to one half an order of magnitude; i.e., from times 0.3 to times 3. When doing such calculations the proximate relationship of 3 x 3 = 10 is valid since it is usually better than some of the estimations used elsewhere. On the other hand, there are instances where a factor of two can lead to a difference of more than one half of an order of magnitude; eg. when a parameter is squared since 2 x 2 = 4. In such instances the Nomograms presented later are quite useful in rapidly assessing the sensitivity of the answer to the parameter assumed at the beginning.

9.1.1 Approximating Capacitance

The capacitance of a free standing sphere is given by Equation 25.

$$C = 4\pi\epsilon\epsilon_o r \qquad (25)$$

$C \equiv$ Capacitance, F
$\epsilon\epsilon_o \equiv$ Permittivity, F/m
$r \equiv$ Radius of the sphere, m

A simple substitution of the value for the permittivity for air (8.85×10^{-12}) into this expression yields the useful rule of thumb that the capacitance of a free standing conductive sphere in picofarads is numerically equivalent to its radius in centimeters. However, as a sphere is brought near a ground plane the capacitance of the sphere increases and approaches infinity as the distance between the surface of the sphere and the ground plane approaches zero. If the ground plane is conductive a spark discharge will occur when the separation distance becomes small. It can be estimated that the spark occurs when the capacitance has doubled. This leads to the rule of thumb that when a spark occurs the capacitance of an object in picofarads is roughly equivalent to its "diameter" in centimeters.

Likewise, the capacitance of a free standing cylinder is half of its height (Haase, 1977) in centimeters and the capacitance of a cylinder to a ground plane when a spark occurs is its "height" in centimeters. The capacitance approximations can be made by eyeballing an equivalent "diameter" or "height" for the object in question. These rules of thumb are quite useful in the estimation of the capacitance of objects common to the workplace as long as the length to width ratio is not excessive. For free standing rods and pipes where the length to diameter ratio is much greater than one, a broad brush approximation of the capacitance is ten picofarads per meter of length.

These rules of thumb apply to single objects in air. If the object is in another medium, the capacitance should be multiplied by the dielectric constant of the surrounding medium.

Also, these rules of thumb cannot be used when there is a large area between two conductive objects which are close together. For instance the case of a ball valve where the ball is isolated by Teflon[tm] seats. The capacitance of the ball is many times the free standing capacitance and in-process measurements may be in order or other estimations may be considered, like assuming some sort of pseudo-equivalent geometry, such as parallel planes.

The capacitance between two finite parallel planes is given by Equation 26. (Neglecting fringe effects.)

$$C = \epsilon\epsilon_o A/\ell \tag{26}$$

C ≡ Capacitance, F
$\epsilon\epsilon_o$ ≡ Permittivity of the medium between the planes, F/m
A ≡ Area of the planes, m^2
ℓ ≡ Distance between the planes, m

For instance, the relation for the capacitance between two concentric cylinders is given by

$$C = 2\pi\epsilon\epsilon_0 h/(\ln a/b)$$

where **h** is the height, **a** and **b** are the radii, with **a** > **b**. Taking the ideal case where ϵ = 1, h = 1 meter, a = 0.052 meter, and b = 0.050 (the outer being 2 mm larger than the inner) one calculates a capacitance of 1.42 nF. But, lets say that the formula for the concentric cylinders was not handy when an approximation was needed. Taking the relationship for the parallel plate capacitor and "rolling open" the ideal cylinder to make one having the length of one meter, a width of 2πa, and a separation of 0.002 meters, one calculates a capacitance of 1.39 nF. A negligible difference for the purposes of approximation.

Note that the capacitance approaches infinity as the distance between the conductors approaches zero and a small difference in separation can make a big difference in capacitance.

9.1.2 Approximating Resistance

The only way to determine with confidence the electrical resistance through an object is to make a measurement with the appropriate equipment. However, for a first cut approximation the resistance of an object can sometimes be surmised by analogy to a similar material. If a ballpark resistivity can be inferred, then a ballpark resistance can follow. Unfortunately the handbooks are of limited utility in finding electrical property data which are specific to one's problem, but there are generic data which can sometimes be quite useful, Tables 6.1 and 6.2.

When an insulative surface is rubbed with a dissimilar material and surface charges are produced, those charges will then dissipate to ground through the path of least resistance. So it is the surface resistivity which is usually of interest because the surface charges find their way to ground across the surface. There are no handy rules of thumb to use for estimating a surface resistivity, but one must bear in mind that humidity plays a very important role, cf., ¶ 6.4.2. The surface resistivity of an insulative material may decrease by a factor of 1×10^6 in going from 30% to 90% relative humidity, and little or no electrostatic effects occur on surfaces at relative humidities greater than 75-80% except on surfaces which are markedly hydrophobic. (Gibson and Lloyd, 1963) Table 6.1 offers some guidance in this regard for surfaces at low humidities.

9.1.3 Approximating Charge

Charge is accumulated on materials as they move through process equipment. In the case of liquids, empirical equations have been derived (eg., Schön, 1962 and Britton, 1988) for the streaming current and the charge density in a liquid as it moves through a pipe. There is enough similarity among insulative liquids to permit the use of some generic relationships for streaming current and charge density, but in the case of solids there are no empirical relationships which can be used with confidence over the variety of materials and process conditions one is likely to encounter in practice.

Extensive studies on streaming currents in liquids have been performed (Britton, 1988) for drum filling operations. It was found that the traditional Schön relationships (Schön, 1962) did not sufficiently embrace some practical situations of liquid flow through smooth-bore pipes. Britton therefore suggests more conservative relationships (greater quantities) for streaming currents and liquid charge densities, ¶ 6.1.1. Britton's determinations are approximately one order of magnitude greater than those of Schön.

Table 9.1: Typical Electrical Properties for Selected Materials

Material	ρ, Ω·m	λ, Ω	ϵ	E_b, MV/m
Metals	10^{-8} - 10^{-5} [a]	-	∞	-
Soil	10^{2} - 10^{4} [a]	-	-	-
Water (extremely pure)	10^{6} [b]		80.4 [b]	-
Trinitrotoluene, TNT	9 x 10^{8} [c]	-	5.2 [c]	3.7
Wood (dry)	10^{8} - 10^{11} [a]	10^{12} [a]		
Concrete	~10^{7} [d]	10^{7} [a]		
Neoprene	10^{11} [e]		9.0 [e]	6 - 24 [e]
Rubber, hard	3 x 10^{11} [f]		3 [f]	0.4 - 0.6 [f]
Rubber, soft vulcanized	1-15 x 10^{13} [f]		2.7 [f]	
Polyurethane, PU	10^{9} - 10^{12} [a]	10^{12} - 10^{15} [a]	5 - 8 [e]	14 - 21 [e]
Glass	10^{9} - 10^{12} [a]	10^{10} - 10^{12} [a]		
Pyrex	10^{12} [f]		4.8 [f]	134 [f]
Polymethylmethacrylate, PMMA	10^{11} - 10^{14} [a]	10^{12} - 10^{16} [a]	3.3 [e]	16 - 20 [e]
Polystyrene	10^{10} - 10^{19} [a]	>10^{14} [a]	2.5 [e]	20 - 28 [e]
Polyvinylchloride, PVC (rigid)	10^{13} [a]	10^{12} - 10^{13} [a]	3.2 [e]	20 - 40 [e]
Polyethylene, PE	10^{14} - 10^{18} [a]	>10^{12} [a]	2.3 [e]	18 - 40 [e]
Polytetraflouroethylene, PTFE	10^{15} - 10^{19} [a]	10^{12} - 10^{17} [a]	2.0 [e]	20 [e]
Air	-	-	1.0 [d]	3 [d]

ρ ≡ Volume resistivity. λ ≡ Surface resistivity. ϵ ≡ Dielectric constant
E_b ≡ Dielectric breakdown strength

References: [a] Walmsley, 1992 [b] Britton, 1992 [c] Fedoroff and Shefield, 1972 [d] Cross, 1987 [e] Rodreguez, 1970 [f] Perry, 1950

Table 9.2: Typical Leakage Resistance Values
(Resistance to earth unless otherwise specified)

Footwear	Leakage Resistance, Ω
Static Dissipative Footwear	$10^4 - 10^8$ [a]
Insulative Footwear	$10^8 - 10^{15}$ [a]
Leather	$10^3 - 10^9$ [b]
Rubber	$>10^9$ [b]
"Conducting Rubber"	$10^4 - 10^7$ [b]
"Microcellular Rubber"	$> 10^{13}$ [b]
ANSI Z41, Type 1	$< 5 \times 10^5$ [c]
ANSI Z41, Type 2	$< 10^4$ [c]
ANSI Z41, Electrical Hazard Footwear: < 5.0 mA at 14,000 V (i.e., $> 2.8 \times 10^6$ Ω) [c]	

Other Selected Items	Leakage Resistance, Ω	
Human Skin, dry	10^4 [a,d]	
Human Skin, wet	$\leq 10^2$ [a]	$\approx 10^3$ [d]
PVC Floor (without additives)	10^{11} [a]	$10^9 - 10^{11}$ [d]
Concrete Floor, dry	10^6 [a,d]	2×10^6 [e]
Wooden Floor, untreated, dry	10^7 [a,d]	
Asphalt Floor	10^{12} [d]	$\geq 10^8$ [e]
Brick, Tile (etc.) Floor	$10^8 - 10^{10}$ [d]	
Linoleum	$10^8 - 10^{12}$ [d]	
Tires	$10^3 - 10^8$ [d,e]	
Between metal objects via oil film (e.g., in bearings)	$\approx 10^3$ [d]	
Flange on glass line	$10^{11} - 10^{13}$ [a,d]	

References: [a] ESCIS, 1988 [b] Brundrett, 1977 [c] ANSI Z41, 1983 [d] Walmsley, 1992 [e] von Pidoll, 1996

The relationships of ¶ 6.1.1 can be used to approximate the amount of electrostatic charge generated in some practical situations. For typical estimates one may want to use the relationships of Schön while for conservative estimates the relationships of Britton may be more appropriate. To this end, Nomogram 9.1 based on the relationships of Britton is useful in approximating streaming currents and charge densities when the flow parameters of the liquid are known; however, one must bear in mind the following caveats:

- Rough bore hoses with inner grounding spirals give streaming currents a factor of 20 higher than smooth-bore chemical hoses. (Britton, 1988)

- Impediments to flow, such as filters, can seriously increase the amount of charge generated. Charge densities of 1,600 $\mu C/m^3$ have been measured (Britton, 1988), and this may be considered as a practical upper limit of charge density in liquid flow situations even though Gavis and Wagner (1968) have observed transient (not steady state) charge densities of 1,900 $\mu C/m^3$.

- In instances where a liquid contains a second phase (i.e., a solid or a gas) the relationships do not apply. Substantial enhancement of charge generation will occur when a second phase is present. (Cross, 1987)

- Stirring produces the same order of magnitude of charging as pipe flow. (Cross, 1987).

In powders the charge is bound to the surface of the particles. The amount of charge which can be impressed upon medium resistivity powders, those having volume resistivities in the approximate range of 10^6 Ω·m to 10^9 Ω·m (BS 5958), have been determined for typical powder operations, Table 2.1. Experience has shown that high surface charge densities of 10^{-7} to 10^{-5} C/m^2 are reached on the particles, but at charge-to-mass ratios above 3×10^{-6} C/kg particles no longer compact because of electric field effects, (Lüttgens and Glor, 1989). The amount of charge on a powder coming from a process unit cannot be packed into heap of the powder because of charge compaction and bulking brush discharge, qv.

The amount of charge which can be accumulated on a heap of powder is one consideration; the time it takes for that charge to dissipate is another. If something is known of a powder's bulk resistivity (or conductivity), then a relaxation time can be reckoned from Equations 3, and 10.

$$\rho = 1/\kappa, \qquad \tau = \rho\epsilon\epsilon_0, \qquad \tau = \epsilon\epsilon_0/\kappa$$

ρ ≡ resistivity, Ω·m
κ ≡ conductivity, S/m
τ ≡ relaxation time, s

High resistivity powders, resistivities above 10^9 Ωm (BS 5958), retain charge for long periods. Not only are their volume resistivities very high, but they are usually hydrophobic so that the surface resistivity of the particles can be quite high as well. Such high resistivity powders can retain a charge for hours or even days when in contact with grounded metal, (Gibson and Lloyd, 1963).

A good appraisal of a powder's tendency to gain a charge and an approximation of its resistivity can be determined by the use of a field test meter. As a powder is being piled up somewhere in a process, as in a tote bin for example, the magnitude of the charge can be assessed by "looking at it" with a field test meter (cf., ¶ 8.4). By monitoring the rate at which the charge dissipates from a quiescent pile sitting on a grounded surface, a relaxation time can be reckoned from Equation 8 - and thus an apparent resistivity from Equation 10.

$$Q = Q_0 \exp(-t/\tau) \quad \text{and} \quad \tau = \rho\epsilon\epsilon_0$$

9.2 Examples of Approximations

9.2.1 Refuelling an Automobile

There was an incident where a person ignited gasoline vapors while refuelling an automobile. At a self-service station he had inserted the nozzle into the tank and began filling by latching the nozzle open. He got back into the drivers seat to retrieve something from his glove box. He then slid across the seat, got back out of the car, and reached for the nozzle. A spark occurred at the fillpipe when he touched the nozzle and the gasoline vapors were ignited.

The electrostatic scenario here is that charge was generated as he slides across the seat, charge was accumulated on his body, discharge occurred between his body and the grounded nozzle, and there are flammable gasoline vapors at the nozzle and the fillpipe.

It is informative to examine whether or not such an electrostatic scenario is likely and what conditions must be in place for it to happen. First, consider the charge generated on the seat of his britches when he slid across the seat. Assume an area some 30 by 30 centimeters or 0.1 m^2. (A quantity which some folks might take a bit of umbrage to, but nevertheless a reasonable estimate.) On a day of low humidity one may even hear a crackle as one slides across a seat, depending upon the nature of the fabrics. This would indicate that the maximum charge density of 2.7×10^{-5} C/m^2 (cf., ¶ 3.2.1) has been reached. This would equate then to a maximum charge on his britches at some 3×10^{-6} C (3 μC). It would be unusual to achieve such a charge but we can run with this estimate while keeping in mind that it is an upper limit.

The capacitance of the human body is taken to be some 200 pF; therefore from Equation 22 or from Nomogram 9.2 we estimate a spark discharge energy of some 20 mJ as an upper limit. The MIE of petroleum vapors is taken to be 0.25 mJ (API RP2003) so this is more than enough to ignite gasoline vapors, even considering the fact that human discharge is less efficient at igniting flammable vapors than is a purely capacitive discharge (Berkey et.al., 1988).

The next consideration is that of the persons footwear. The amount of charge reckoned above is the charge which was on his body the instant his britches parted from the car seat. The charge would then have begun to dissipate through his shoes, and depending upon the total resistance to ground it may have dissipated before he could have reached the nozzle. Again assume a worst case condition and draw another line across Nomogram 9.2 through 0.25 mJ and 200 pF to obtain an estimate of the amount of charge which must be on his body in order to obtain an ignition of the gasoline vapors; i.e., some 0.3 μC.

Assume that it takes 3 seconds for him to reach the nozzle. What must the resistance through his shoes be in order to dissipate the charge below the critical value of 0.3 μC? This can be reckoned from Equations 8 and 9.

$$Q = Q_0 \exp(-t/\tau) \qquad \tau = RC$$

Q ≡ Charge, C; at time t, s [0.3 μC at 3 s]
Q_0 ≡ Initial charge, C [3 μC]
τ ≡ Relaxation time, s
R ≡ Resistance, Ω
C ≡ Capacitance, C [200 pF]

One can solve these relations for the time required for the charge to reach 0.3 μC given that it started at a charge of 3 μC.

$$Q/Q_0 = (0.3/3) = 0.1 = \exp(-3/RC)$$

The natural logarithm of both sides of this equation gives the relation

$$-2.3 = -3/RC \quad \text{or} \quad RC \approx 1$$

So with a capacitance of 200 pF his resistance to ground must be less than $1/(200 \times 10^{-12})$ or $< 5 \times 10^{9}\ \Omega$ for the charge to dissipate in some 3 seconds..

Now, assuming that the pavement was concrete and he had on shoes with neoprene soles 3 mm thick. Was the resistance to ground less than $5 \times 10^{9}\ \Omega$? Referring to Table 6.1, a typical resistivity of neoprene is $10^{11}\ \Omega \cdot m$, and assuming an area on the bottom of his foot as something like 3 cm by 10 cm or 0.03 m^2. Then the resistance to ground through his shoe would be given by Eq. 2.

$$R = \rho \ell / A$$

R ≡ Resistance of conductor, Ω
ρ ≡ Resistivity of conductor, $\Omega \cdot m$
ℓ ≡ Length of conductor, m
A ≡ Area of conductor, m^2

$$R = (10^{11})(0.003)/(0.03) = 10^{10}\ \Omega$$

Not small enough to prevent the incident, but close.

Referring to Table 6.2 one can see that if he had had on static dissipative footwear, the charge would have dissipated through his shoes and the concrete before he reached the nozzle (assuming the above assumptions are reasonable for the specific instance). On the other hand if the pavement had been dry asphalt, the charge may not have found its way to ground.

9.2.2 Filling a Gasoline Can

There have been several incidents during the filling of gasoline cans in the beds of pickup trucks with plastic bed liners. As the cans are filled, the gasoline vapors are ignited. The vapors usually continue to burn around the mouth of the can but are usually extinguished before any real damage is incurred.

The electrostatic scenario here is that charge is generated by the flow of gasoline through the hose, the charge is accumulated on the isolated gasoline can, spark discharge occurs between the mouth of the gasoline can and the grounded nozzle, and there are flammable vapors at the mouth of the can.

It is informative to examine whether or not such an electrostatic scenario is likely and what conditions must be in place for it to happen. First, consider the streaming current in the gasoline as it flows through the hose. Assuming a flow rate, F, of 10 gallons per minute through a ~½" diameter, d, hose; Nomogram 9.1 (or Equation 18) shows that a streaming current of 0.1 μA can be expected under optimum conditions. Assume that the can was a 5 gallon jerry can and ignition occurred when the can was one half to two thirds full. At 10 gallons per minute this equates to some 15 to 20 seconds of fuel flow. At a streaming current of 0.1 μA for 20 seconds there will be 2 μC of charge put into the gasoline can. Assuming that the can has a capacitance of 30 pF (cf. ¶ 9.1.1), Nomogram 9.2 (or Equation 15) shows that a discharge energy of some 70 mJ is to be expected. More than enough to ignite petroleum vapors with a MIE of 0.25 mJ, (API RP2003).

This approximation was made using the conservative equations of Britton, cf. ¶ 6.1.1. Perhaps a more realistic assessment should be made using the equations of Schön, ¶ 6.1.1, which are an order of magnitude less than those of Britton. Therefore, if one assumes a streaming current of 0.01 μA to give a charge of 0.2 μC, Nomogram 9.2 (or Equation 15) shows a discharge energy of 0.7 mJ;. again, enough to ignite petroleum vapors. (Note that a difference in the estimate of one order of magnitude in charge results in a difference of two orders of magnitude in spark energy.)

The other facet which must be considered in this scenario is the isolation of the gasoline can. Pickup truck bed liners are made of polyethylene with a carbon black filler. Table 6.1 suggests that an "unfilled" polyethylene would have a volume resistivity of some 10^{18} Ω·m. The addition of carbon black will reduce the volume resistivity but it will also degrade the physical properties of the polymer. Manufacturers put enough carbon black into the formulation to make the polyethylene black, but not enough to make it "conductive". Therefore, a volume resistivity of some 10^{14} Ω·m may not be unreasonable. Also, the dielectric constant for the polyethylene can be taken as 2.3. The relaxation time of the charge on the gasoline can then be reckoned from Equation 10.

$$\tau = \rho \epsilon \epsilon_0$$

τ ≡ Relaxation time, s
ρ ≡ Resistivity, Ω·m (10^{14})
ϵ ≡ Dielectric constant (2.3)
ϵ_0 ≡ Permittivity of a vacuum, F/m (8.85×10^{-12})

$$\tau = (10^{14})\,(2.3)\,(8.85 \times 10^{-12}) \approx 2{,}000 \text{ s}$$

This is much longer than the time required to fill the gasoline can and suggests that the charge will not dissipate through the bedliner.

9.2.3 Flexible Intermediate Bulk Container (FIBC)

When an MIE is determined for a powder the experimental equipment used has an upper limit to the amount of energy which can be delivered to the spark discharge. If a material does not give a positive result when tested at this upper limit, the material is said to have a MIE greater than the upper limit of the test apparatus. Many times it is thought that such materials will not be ignited by a spark discharge and can be treated as being "nonexplosible". On the contrary, there have been instances with such materials, and they should not be considered as materials which cannot form explosible dust suspensions.

There was an operation in which metalized, groundable FIBCs were being used and reused with a powdered material having a MIE of > 5,000 mJ (> 5 J). A question was asked as to the safety of using these FIBCs with such a material. Could it be shown from a worst case analysis that it would be "safe" to use these FIBCs with this material without having the bags grounded? The answer is a resounding "NO", but a worst case analysis, and a "not so bad" case analysis, is informative.

The FIBCs in question were cuboids approximately 3 feet ($\approx$1 m) on an edge and contained approximately 2,200 lb ($\approx$1,000 kg) of powder. When the FIBCs were emptied the powder flowed out of them in a matter of some 30 seconds. As the powder leaves the FIBC it carries a charge with it and leaves the equal and opposite charge on the inside surface of the FIBC. In a worst case scenario, the inside surface of the FIBC would be charged to the maximum surface charge density of 2.7 $\mu C/m^2$, (cf. ¶ 3.2.1). The area of the inside surface of the FIBC is some 6 m^2; therefore there will be a charge of $(3 \times 10^{-5}) \times (6) \sim 10^{-4}$ C or 100 μC on the inside surface of the FIBC! Assuming the capacitance of the FIBC at 100 pF (cf. ¶ 9.1.1), Nomogram 9.2 shows that a spark discharge energy far in excess of 10,000 mJ would be expected. The scale of the Nomogram is exceeded but a calculation shows $\sim 5 \times 10^5$ J! This worst case estimation shows that there would be an extreme hazard to shocking personnel independent of the explosive nature of the material.

Another approach is to take the data shown in Table 2.1 for a powder sliding down an incline. The datum of 10 $\mu C/kg$ would indicate that there would again be some 100 μC of charge separation, and lead to the same conclusion.

Realistically speaking these estimates are upper limits which could not be reached in actual practice, but it is informative to work the problem the other way. The datum of MIE > 5,000 mJ can be used as the starting point and working backwards. Nomogram 9.2 shows that a spark discharge energy of 5,000 mJ from a 100 pF capacitor would require a charge on the capacitor of some 30 μC. So 30 μC on 1,000 kg would equate to a charge density of some 0.03 $\mu C/kg$ on the powder coming out of the FIBC to leave the equal and opposite surface charge on the bag at a surface charge density of $30/6 = 5\ \mu C/m^2$. These charge densities are within the range of those observed in everyday powder operations; therefore, highly energetic discharges would be expected if metalized FIBCs are used and reused without being properly grounded.

All of this goes to show again that ALL CONDUCTIVE OBJECTS IN MATERIAL HANDLING OPERATIONS MUST ALWAYS BE KEPT GROUNDED.

9.2.4 The Minimum Capacitor for Incendive Discharge

In the analysis of some practical situations the question sometimes arises as to how big a metallic object must be in order to have an incendive discharge. There may be cases where a small metallic object, such as a bolt or rivet, may be electrically isolated and the concern is that it may act as an accumulator which could lead to an incendive discharge. An examination of Equations 13, 14, and 15 and Nomogram 9.2 suggests that for a given spark discharge energy, the potential on an object must be quite large for a small capacitor. It is therefore intuitively attractive to say that there are objects which are too small to constitute a problem. It would be convenient to put this notion into some perspective. This can be done by considering some theoretical limit.

There is no doubt that charges can be accumulated or induced upon metallic objects and that there can be capacitive discharges from them. Considering the "ordinary" flammable vapors which have MIEs of 0.25 mJ, the capacitance of a metallic object must be large enough to store this amount of energy. One can then derive a theoretical lower limit to the size (capacitance) of an object because there is an upper theoretical limit to the potential which can be impressed upon a free standing metallic object.

Assume the object to be a sphere. The amount of charge which can be impressed upon the sphere will be limited by the maximum surface charge density in air as derived from Equation 11 where E_{max} is 3×10^6 V/m.

$$\sigma_{max} = \epsilon\epsilon_0 E_{max}$$

It follows then that the maximum charge which can be impressed upon a sphere is the maximum charge density times the surface of the sphere.

$$Q_{max} = 4\pi\sigma_{max}r^2 = 4\pi\epsilon\epsilon_0 E_{max}r^2$$

This maximum charge corresponds to the maximum potential as given by Equation 5.

$$V_{max} = Q_{max}/4\pi\epsilon\epsilon_0 r$$

The maximum potential on the sphere can then be expressed in terms of the maximum field strength by combining the previous two expressions.

$$V_{max} = E_{max}r$$

Turning now to Equation 13, it can be rearranged to reckon the capacitance which will hold 0.25 mJ at the maximum potential.

$$C = 2W/V_{max}^2 = 2W/(E_{max}r)^2$$

The capacitance of a free standing sphere is given by Equation 25.

$$C = 4\pi\epsilon\epsilon_0 r$$

The previous two expressions then yield the relation which gives the radius for the minimum sphere which can hold 0.25 mJ.

$$r^3 = W/2\pi\epsilon\epsilon_0 E_{max}^2$$

$$r = \sqrt[3]{\frac{0.00025}{2\pi(8.85 \times 10^{-12})(3 \times 10^6)^2}}$$

$$r = 0.008 \text{ m} = 8 \text{ mm}$$

This theoretical result suggests that there can be incendive discharges from objects as small as ~5/8" in diameter, but one must consider the context: [1] the object is spherical, real objects sometimes have sharp corners where corona discharge would occur, [2] the object has a maximum surface charge which could only occur in cases of very intense electrification where there would be other considerations, [3] the capacitance of real objects would be different, and [4] real objects are seldom if ever isolated they have some resistance to ground.

In practical situations it is known that a beer can is big enough to accumulate enough charge to ignite ordinary flammable vapors; but how small is too small is an open question.

One straight line through the scales simultaneously solves the relations

$I_s = 2.5 \times 10^{-5} \upsilon^2 d^2$ $\quad S = 3.18 \times 10^{-5} \upsilon$ $\quad F = (\pi/4) \upsilon d^2$

(To convert from ft/sec to m/s multiply by 3.28)
(To convert from bbl/hr to m^3/hr multiply by 0.159)

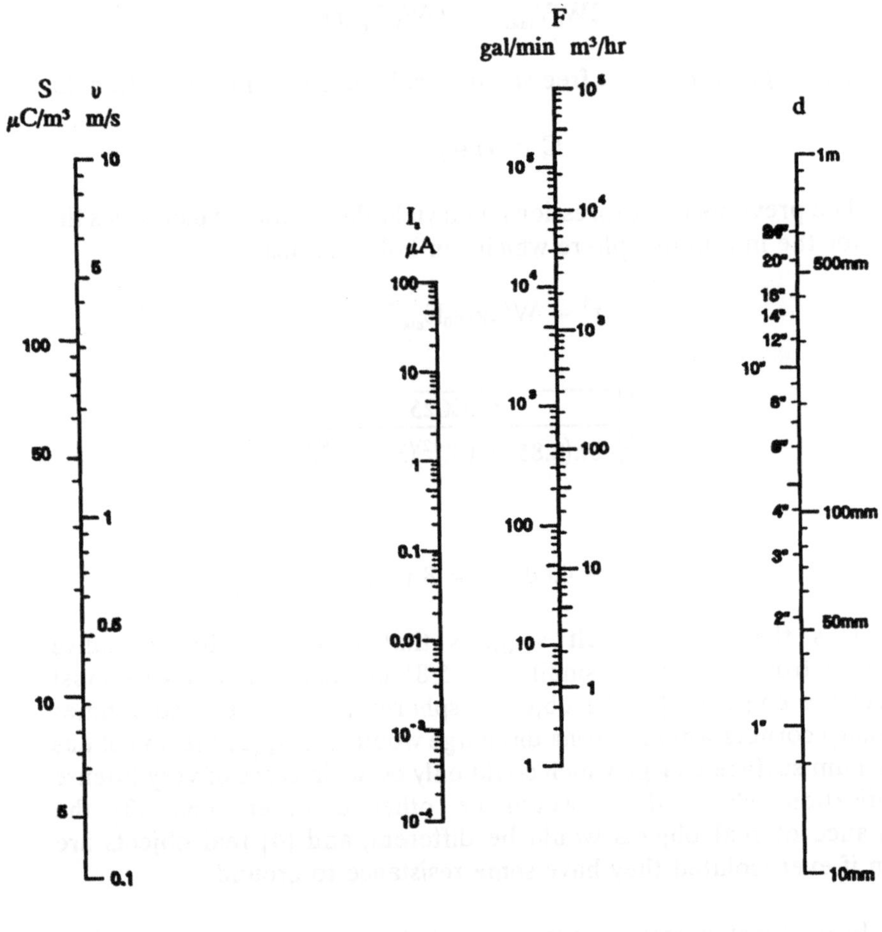

Nomogram 9.1: Nomogram for Estimating Charge on Insulative Liquids while Flowing through Long, Smooth Bore Pipes.

One straight line through the scales simultaneously solves the relationships

$W = ½CV^2$ $\qquad$ $W = ½QV$ $\qquad$ $W = Q^2/2C$ $\qquad$ $Q = CV$

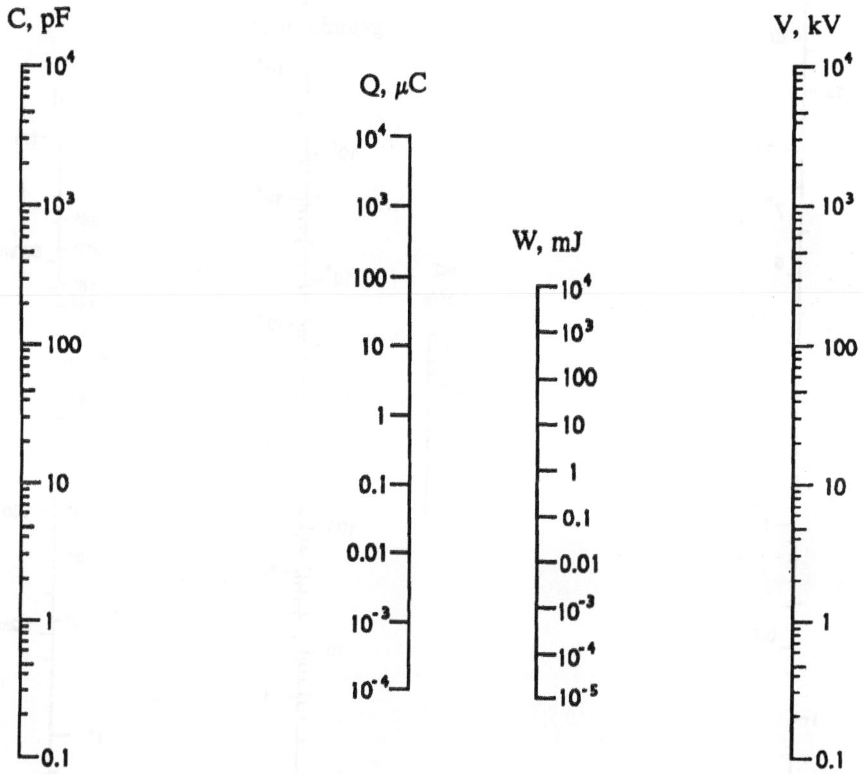

Nomogram 9.2: Nomogram for Estimating the Energy in a Capacitive Spark Discharge

One straight line through the scales solves the flow relation

$$F = (\pi/4)\upsilon d^2$$

(To convert from bbl/hr to m^3/hr multiply by 0.159)

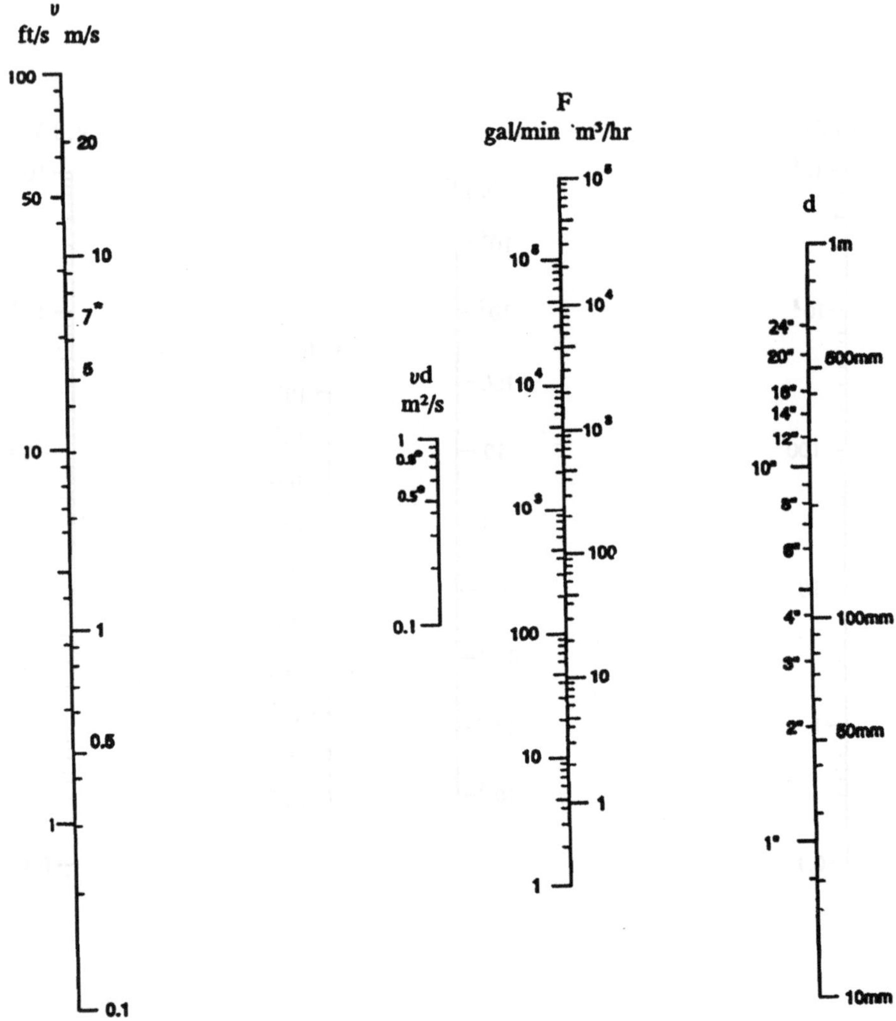

Nomogram 9.3: Nomogram for Estimating Fluid Flow Parameters

Chapter 10

Case Histories

Perhaps the best way to obtain a feeling for electrostatic scenarios is the examination of case histories. To this end the following case histories are presented. During the investigation of an incident where electrostatics is suspected to be the ignition source, it is always incumbent upon the investigator to quantify the proposed scenario as much as possible. As can be seen below, there are cases where assumptions must be made to fill in the gaps between the facts; nevertheless, the scenarios must in the end be reasonable. Many times they are evident, but the quantification should demonstrate the postulated scenario as much as possible.

On the other hand there are times when quantification is difficult at best and an electrostatic scenario must be postulated without quantification. Nevertheless, in these instances the explanation must be in accord with established electrostatic mechanisms .

10.1 Vacuum Truck Emptying a Sump

There was a below grade sump containing some off specification liquid which primarily consisted of toluene. A vacuum truck was summoned to remove the liquid. The operator failed to ground the truck and proceeded to empty the open sump. When the sump was almost empty, the vapors in the sump were ignited.

Process conditions:

Liquid:

Conductivity: 1×10^{-10} S/m (measured)
Dielectric constant: 2.4 (Table 6.1d)
Flash point: ~ 4.4°C (40°F)

Truck:

Volume: 54.6 m^3 (12,000 gal.)
Pumping rate: 0.315 m/s (500 gpm.)
Pipe diameter: 0.102 m (4 in.)
(Velocity: 3.85 m/s)
Capacitance: ~ 1000 pF
Resistance to ground: unknown

Ambient temperature: 25°C (77°F)

The streaming current for pumping hydrocarbons at these rates is given by Equation 16 or Nomogram 9.1 which is some 3 μA. This represents a near maximum condition for the pumping of hydrocarbons and does not consider two phase flow. But, ignition occurred when the sump was almost empty; therefore, it is assumed that there was two phase flow in the hose because air was being entrained into the flow of the liquid; i.e., the "soda straw effect". For a first cut, it is assumed that there was a streaming current of 3 μA.

This is a case of a constant amperage source of ¶ 2.10.2. A streaming current going into the vacuum truck which puts an electrostatic charge on the truck. There is a simultaneous discharge current flowing to ground through the tires to the pavement, and the amount of charge residing on the truck will be determined by the resistance of the truck's body to ground. The voltage on the truck is given by Ohm's Law, Equation 1.

$$V = I_s R$$

where

V is the voltage on the truck
I_s is the streaming current, and
R is the resistance of the truck to ground.

But R is unknown; therefore, one must assume some reasonable values for the resistance of the truck to ground. An ungrounded tank truck on rubber tires could have a resistance to ground as low as 10^7 Ω if it were on concrete (cf. Table 9.1) or as high as 10^{12} Ω if it were sitting on dry asphalt (cf. Table 9.2). Then from Ohm's Law a table of possible voltages can be constructed.

Resistance to Ground, Ω	Voltage on Truck, V
1 x 10^{12}	3,000,000
1 x 10^{11}	300,000
1 x 10^{10}	30,000
1 x 10^{9}	3,000
1 x 10^{8}	300
1 x 10^{7}	30

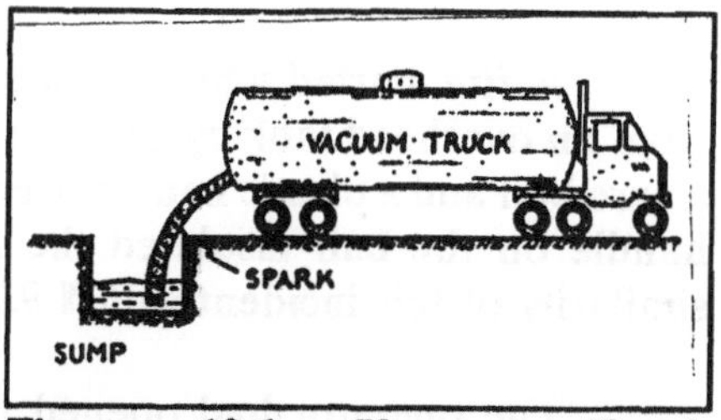

Figure 10.1: Vacuum Truck Emptying a Sump

The truck was sitting on asphalt but it is doubtful that a potential of 3 MV is possible because there would be sparking somewhere in the system before such charge buildup could occur. Since the ignition of the vapors in the sump actually occurred, it is postulated that the spark was between the spiral wire in the hose to the metal lip of the sump, Figure 10.1. This mean's that the spark went through some 2 mm of rubber and had enough pizazz left over to cause a discharge in a air gap which exceeded the MIE of the toluene.

The breakdown of neoprene is some 10 MV/m (Table 9.1), so breakdown across 2 mm will occur at some 20,000 V. Therefore, if the resistance to ground of the truck is some 10^{10} Ω, there would be repetitive sparking from the spiral wire to the lip of the sump. If the spark must also jump across an air gap to reach ground, then any vapors which may be present would be ignited.

It is postulated that in the actual case the resistance to ground of the truck was marginal and that there were no incendive sparks during the time of the pumping of the hydrocarbon. When two phase flow began (the "soda straw" effect) there was a significant increase in the streaming current and the incendive spark postulated above occurred. This leads to the conclusion that sparking occurred from the truck when it was at a potential near 20,000 V. The energy in the spark can then be determined from Equation 13 or Nomogram 9.2.

$$W = \frac{1}{2}CV^2 = (0.5)(1{,}000 \times 10^{-12})(20{,}000)^2 = 0.2 \text{ J}$$

This is three orders of magnitude more than enough to ignite toluene alone.

10.2 Drawing Toluene into an Ungrounded Bucket

A fire started when an operator was drawing a bucket of toluene from an overhead tank by gravity flow. He had hung a metal bucket with a wire bail and a plastic handle over a globe valve, Figure 10.2. The plastic handle on the bail insulated the metal bucket from ground. (Note the similarity of this incident with ¶ 9.2.2.)

The operator had opened the valve to draw the toluene and had backed away from the bucket. In a few moments the toluene ignited. The operator then left the scene and quickly returned with a small fire extinguisher, which proved to be inadequate. He then left the scene and returned with a large fire extinguisher, but by the time he had returned the fire was out of control and the bucket was overflowing with burning toluene.

Process conditions:

Equipment:
- Pipe diameter: 0.75" (nominal)
- Valve: 0.75" globe valve (hardware store variety)
- Filter: ~2' upstream of valve
- Capacitance of bucket: 20 pF (measured)

Toluene:
- Conductivity: 3 x10^{-9} S/m (measured)
- Dielectric constant: 2.4
- Flow: 5 gal/min (nominal)

The investigation showed that the operator had opened the valve and backed away from the bucket. The operator said that "I was just standing there looking at it when it caught fire." Therefore, discharge from the operator was ruled out and the scenario of a streaming current was considered.

A streaming current can be reckoned from Equation 16 or Nomograph 9.1 and found to be of the order of 0.01 μA were it not for the presence of the in-line filter. The residence time of the toluene between the filter and the exit of the pipe was less than one second, much longer than the recommended 30 seconds; therefore, a streaming current at the exit of the pipe of 0.1 μA is not an unreasonable estimate. In any case, an estimate for the streaming current can be taken to be between 0.1 μA and

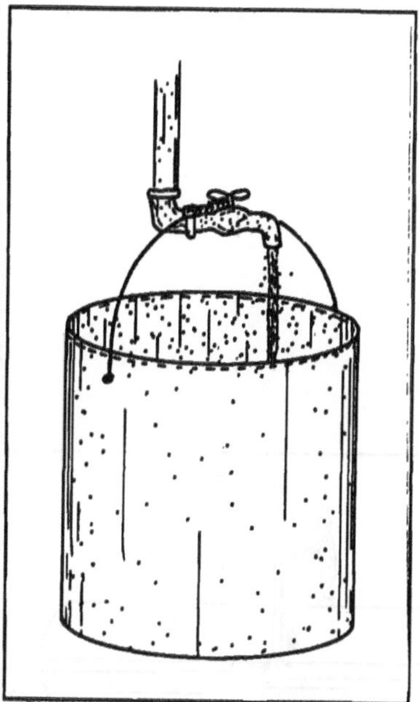

Figure 10.2: Drawing Toluene into an Ungrounded Bucket

0.01 μA. It is interesting to perform the quantification with both of these estimates, beginning with the larger.

Assuming that the flow of the toluene continued for some 30 seconds, then there would have been 3 μC of charge on the bucket provided that the bucket was completely isolated. The energy on the bucket can be found from Equation 15 or Nomogram 9.2:

$$W = Q^2/2C = (3 \times 10^{-6})^2/2(20 \times 10^{-12})$$
$$= 0.225 \text{ J} = 225 \text{ mJ}$$

And the voltage on the bucket can be found from Equation 12 or Nomogram 9.2 as being 150 kV. With the breakdown strength of air at 3×10^6 V/m then a spark from the bucket could jump across a gap of 0.15 m (or some 6"), eg., from the wire of the bail to the body of the globe valve.

On the other hand if the streaming current were 0.01 μA the calculation yields an energy of 2.25 mJ, a voltage of 15 kV, and a spark gap of 0.015 m (or some 5/8").

The MIE of toluene is 0.24 mJ so it can be seen that there is much more than enough for ignition of the vapors, even under the most conservative assumptions.

The lesson relearned here is to keep all metallic equipment well grounded anywhere there can be flammable vapors.

10.3 Sampling while Loading a Rail Car.

An ignition of vapors occurred when an operator attempted to catch a sample during the loading of toluene to a rail car. The explosion damaged some external piping and there was a spill of product and an ensuing fire. The fire brigade brought the fire under control before there was any damage to adjacent units in the refinery.

Process Conditions:

Piping:

Fillpipe: 4 inch
Fill rate: 200 gallons/minute
Rail car: 26,000 gallons

Toluene:

Conductivity: Unknown
Dielectric constant: 2.4
Flash point: 4.4°C (40°F)

Figure 10.3: Sampling a Rail Car while Loading

The pumper had inserted the fillpipe approximately half way into the rail car such that there was splash filling. When the car was approximately half full he lowered a weighted metal sample cage into the ullage in an attempt to catch his sample. The sample cage was lowered by a insulative rope. Ignition of the vapors most probably occurred by a spark discharge between the metal sample cage and the fillpipe, Figure 10.3. The charging mechanism was either splash filling or streaming current, or some of both. The charge from splash filling accumulated on the mist in the ullage. The charge from the streaming current accumulated on the liquid in the tank. These charges could well have been of the same polarity, in which case they would have been additive. The metal sample cage acted as an electrostatic accumulator by the induction of the space charge onto the cage. As the cage was lowered into the ullage the amount of induced charge increased; the deeper it went the more the induced charge. When the cage approached the grounded fillpipe, the spark occurred and ignition of the vapors was accomplished.

During the investigation, several procedural errors were found. Several loading stations at the refinery were equipped with fill-pipe assemblies for the top dome loading of rail cars. The fillpipes were articulated such that they could be swung over the dome and inserted into the rail cars to any desired depth. The pumpers knew by experience that if the fillpipes were not lowered into the rail cars enough, liquid would splash out of the dome. Therefore, they had become accustomed to lowering the fillpipes into the tanks enough to prevent the liquid from being splashed on the outside of the rail car, but did not always take care to see that the fillpipes were lowered "all the way" so that internal splashing would be minimized. After the incident it was established that most of the time the fillpipes were lowered approximately half way to the bottom. Thus, there was splash loading for an appreciable period at the beginning of a load.

The written procedure stipulated that the rail cars were to be sampled after loading was completed, not during loading. Nevertheless, it was operationally convenient for the pumpers to have the samples ready when the courier came by on his regular schedule. This convenience led to the practice of taking samples at any time during the loading process so that the samples would be available to the courier when he made his rounds. Otherwise, if the pumper waited until filling was complete, the sample may not get to the laboratory before the rail car was dispatched. Thus, catching samples during loading became a common and had been practiced for a year or two before the incident.

In such instances the question is always asked as to why there had been no previous problems since "we have always done it that way". The answer lies in the fact that several things have to be "just right" in order to get ignition.

There were some 20 different liquids which could be transferred at any of the loading stations. All but toluene were either high vapor pressure products which formed fuel rich vapors or low vapor pressure products which formed lean vapors. Toluene was the only one at this location which would form explosive vapors at ambient temperatures. That is, the other liquids were somewhat forgiving about the formation of flammable vapors.

There had to be enough space charge in the tank to induce itself on the sample cage. If the charged mist were the culprit, the position of the fillpipe had to be such that splashing was optimized. If the charged liquid were the culprit, then the conductivity of the toluene had to be within certain limits (cf. ¶ 6.1.1). It is known that the conductivity of a product such as toluene can vary by one or two orders of magnitude from batch to batch. As was the practice at the refinery, there was no conductivity additive in the toluene; therefore, the conductivity of the toluene on that day was high enough for a big enough streaming current but low enough to retain a charge.

The geometry of the sample cage as it touched the fillpipe had to be within certain limits in order to effect ignition. Had it touched the grounded fillpipe many times on the way down, the induced charge would not have built up. On the other hand if it swung away until it was just above the liquid and then swung over to the fillpipe, then the spark would have been optimized. If it got below the surface of the liquid before touching, then the spark would not have been in the vapors

As in many other electrostatic scenarios there are several things which must come together just right in order to have ignition. The lesson relearned here is not to perform simultaneous operations, cf. ¶ 7.3.

There are the alternate ignition scenarios of [1] a brush discharge at the metal sample catcher, [2] a brush discharge at the fill pipe as the liquid level of the charged toluene rose, and [3] a spark discharge between the tank wall and a piece of trash floating on top of the liquid. But, in any case, it is concluded that the explosion was an ignition of toluene vapors by an electrostatic discharge.

10.4 A Vapor Ignition in a Road Tanker, I

An ignition of gasoline vapors from a previous load occurred when an operator was loading diesel oil into an old tank truck. The operator suffered some severe burns when the flames burst forth from the dome.

Process conditions:

Truck:
- Subject compartment: 200 gallons
- Fill rate: 30 - 40 gal/min (nominal)
- Grounded by ground wire

Diesel oil:
- Conductivity: unknown
- Dielectric constant: 2.5

An insurance inspector had made an inspection at a distributor location and recommended that the old compartmented delivery trucks no longer be splash loaded through open domes. It was suggested that an

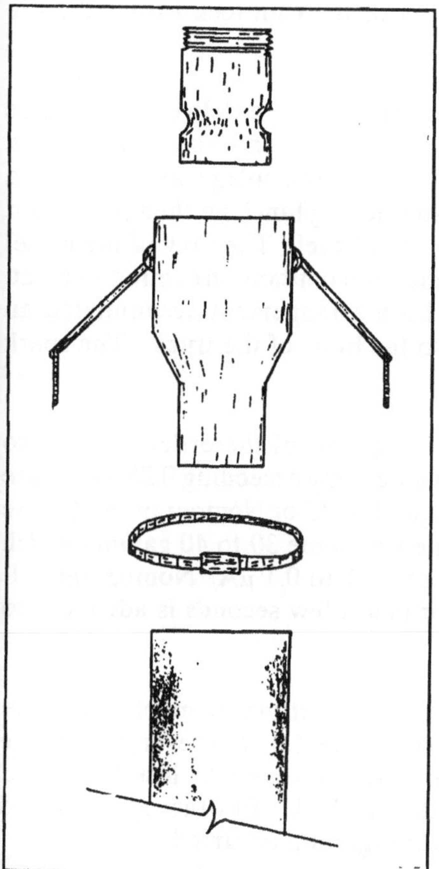

Figure 10.4 Road Tanker I, Hardware Store Items for Modifying a Nozzle

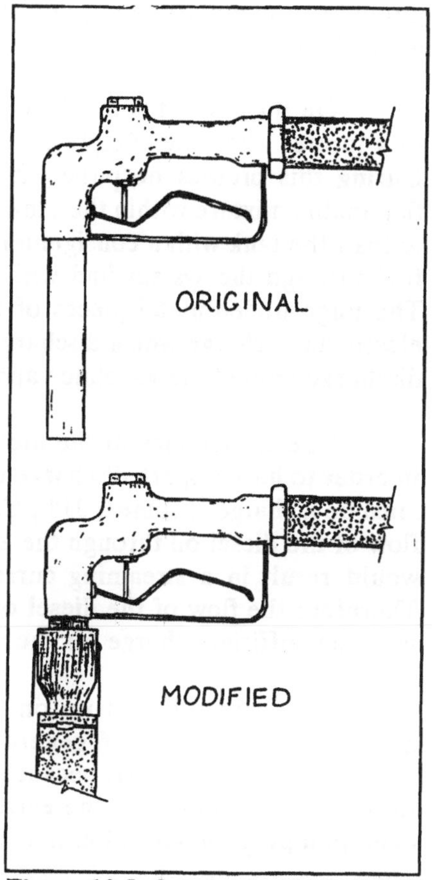

Figure 10.5: Original and Modified Nozzle

extra length of fill pipe be added to the nozzle so that it would reach the bottom of the tank and thus be bottom loaded as the standards suggest. In this manner the operator could load as he had always done through the open dome but the flow of the liquid would enter each tank compartment at the bottom.

The operator went to the local hardware store and purchased a PVC cam lock fitting with metal ears, some insulative hose, and a metal band clamp, Figure 10.4. He then retrofitted the metal, grounded nozzle with the cam lock fitting and connected the hose with the band clamp. The connecting pieces were made of both metal and PVC so that there were some metal components which did not have a metallic path to the ground of the metal nozzle; i.e., the ears of the cam lock fitting and the band clamp, Figure 10.5.

On the day of the incident the operator made a delivery of gasoline to a farmer where he completely drained the tank and hose connections. During this process he pulled fresh air into the ullage and created a flammable mixture within the tank. After having lunch he then proceeded to load the tank with a consignment of diesel fuel. The flow of the diesel fuel through the system had with it the concomitant streaming current. The ungrounded metal pieces of the hose arrangement accumulated an electrostatic charge which discharged to the body of the truck. The spark discharge ignited the gasoline vapors.

The capacitance of the metal fittings was of the order of 3 pF, so in order to have a spark discharge with an energy exceeding 0.25 mJ, there must be a charge of some 0.038 μC, (Equation 15 or Nomogram 9.2). The flow of the diesel oil through the 2" line was some 30 to 40 gal/min which would result in a streaming current of 0.01 to 0.1 μA, Nomogram 9.1. Therefore the flow of the diesel oil for just a few seconds is adequate to separate sufficient charge for the spark.

The charge builds up on the camlock fitting in much the same manner as a Van de Graaff generator, cf. Figure 1.5. However, the charge may build up rather slowly because it must leak through the insulative PVC fittings to accumulate on the ears of the camlock. But the operator had been pumping for several minutes before ignition occurred.

10.5 A Vapor Ignition in a Road Tanker, II

At a distribution location where diesel fuel was gravity fed into delivery trucks, the fuel was splash loaded through the domes of the compartments. An ignition of vapors from a previous load of a volatile material occurred about half way through the loading of a 200 gallon compartment. The operator suffered some second and third degree burns.

Process conditions:

Truck:

Subject compartment: 250 gallons
Fill rate: 20 - 30 gal/min (nominal)
Grounded by ground wire

Diesel oil:

Conductivity: unknown
Dielectric constant: 2.5

Again, this is an example of switch loading and at this location it was most unusual to have a cargo of anything but diesel oil, but there had been a special cargo delivered the previous day as a favor to a good customer.

The normal loading procedure was the gravely feed of diesel oil from an elevated tank to the delivery truck. The operator would insert the nozzle into the dome of the truck, begin the flow, and keep his foot on the nozzle while he watched the liquid level rise until the tank was full. With the nozzle, as purchased from the vendor, it was found that the fillpipe nipple would easily come out of the dome opening if the nozzle arrangement flopped sideways. This was because the nipple fitted into the nozzle housing was too short. It was therefore decided to lengthen the nipple; i.e., replace the 10 inch long nipple with a 16 inch long nipple.

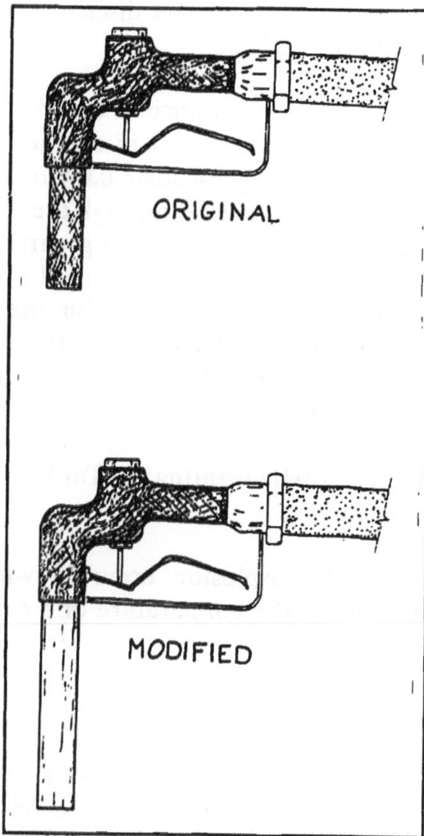

Figure 10.6: Road Tanker II: Original and Modified Nozzle

In this case the original nozzle housing and the nipple were made of PVC as supplied by the nozzle manufacturer. The pipe sizes and threads were standard two inch npt threads, so it was natural that the replacement nipple would be made of a standard section of longer pipe. It was. But the replacement nipple was metal rather than PVC! Thus, the metal replacement nipple was ungrounded because it screwed into the PVC housing of the nozzle, Figure 10.6.

At the time of the incident the flow of the diesel fuel through the system had with it the concomitant streaming current. The ungrounded nipple accumulated an electrostatic charge which discharged to the body of the truck. The spark discharge ignited the solvent vapors left over from the previous load.

The capacitance of the metal nipple was of the order of 10 pF, so in order to have a spark discharge with an energy exceeding 0.25 mJ, there must be a charge of some 0.07 μC, (Equation 15 or Nomogram 9.2). The flow of the diesel oil through the 2" line was some 30 to 40 gal/min which would result in a streaming current of 0.01 to 0.1 μA, Nomogram 9.1. Therefore the flow of the diesel oil for just a few seconds is adequate to separate sufficient charge for the spark. It should be noted that the accumulation of the electrostatic energy on the nipple is much the same as a Van de Graaff generator (cf. Figure 1.5) and it is thus a very efficient electrostatic accumulator.

10.6: Instrumenting a Tank Containing Steam and a Flammable Atmosphere

An explosion occurred when a marine surveyor attempted to determine the temperature of a cargo of 6 oil with a temperature probe.

Process conditions:

Ship length: 711 feet
Deadweight: 48,915 long tons
Subject tank volume: 169,404 cf
Cargo: 6 oil
Previous cargo: condensate
Cargo temperature: unrecorded

A cargo of 6 oil was put aboard the Maltese tank vessel FIONA in Belfast, Ireland for shipment to Northport, New York. Because 6 oil has a high flash point (150-270° F) it was considered to be nonflammable so the master of the vessel did not inert the tanks during the voyage. Since 6 oil has a high viscosity it is necessary to heat the cargo to some 70°C (160°F) in order pump it as a liquid. Near the end of the voyage steam was applied to the heating coils of all of the tanks.

When the pumpman opened the ullage hatch of number one tank to measure the tank ullage, the ullage was full of steam from a leak in the steam coils. It was then decided to shut off the steam but the question remained as to whether or not the cargo was hot enough to pump. As the steam valves were being closed, a surveyor inserted a temperature probe into the tank; i.e., instrumenting while there was still some steam going into the tank. After making his determination, he then began withdrawing the temperature probe. "*Just as the metallic probe reached the level of the hatch cover, the chief mate, who was standing over the ullage opening looking into tank 1, saw a 'yellow light' in the tank....*" The explosion killed the surveyor and injured the pumpman and the chief mate.

It was concluded in the NTSB report that the No. 6 oil contained impurities from the previous cargo of condensate and it was the leftover vapors of condensate which provided the fuel for the explosion. Since the master had not inerted the tanks, atmospheric oxygen (air) provided the oxidizer. The steaming of the tank separated an electrostatic charge which accumulated on the cloud of steam in the ullage. This charge induced itself onto the temperature probe such that there was an incendive brush discharge near the hatch opening; or, alternatively, if the temperature probe was ungrounded and a spark discharge occurred between the temperature probe and the hatch cover.

Getting a quantitative handle on this incident is not straight forward since there are no good rules of thumb for a space charge from a steam generator, other than to know that a lot of charge would be expected. Experience has shown that steaming results in a significant space charge and the insertion of any object into a cloud of steam should always be prohibited if there is *any* chance of having a flammable atmosphere.

10.7: Conductive Liquid in a Plastic Carboy

An ignition of flammable vapors within a mixer occurred when an operator was pouring a liquid additive from a plastic carboy through an open port. The hot gasses from the combustion reaction vented through a properly sized rupture disc into a knockout pot, but a flame also came through the open port to inflict first, second, and a few third degree burns to the operator's face, hands, and forearms.

Process Conditions:

Ambient conditions:
27° C (80° F)
85% RH

Mixer:
Volume: 1.89 m^3 (500 gallons)

Internal temperature: 40° C
Pressure: ambient
Contents: 1.51 m^3 of a proprietary mixture (400 gallons)

Additive:
Temperature: 12° C (54° F)
Flash point: 48° C (120° F)
Conductivity: ~ 10^{-6} pS/m[a]
Dielectric constant: ~ 10[a]
[a] by analogy from literature values

Additive carboy:
Volume: 0.0189 m^3 (5 gallon), "full"
Material: polypropylene
Conductivity: ~ 10^{-16} pS/m
Surface conductivity: unknown

Carboy box:
Material: cardboard with a "plastic" finish
Conductivity: unknown

Figure 10.7: Pouring Liquid into a Mixer from a Carboy

In the makeup of a batch of a proprietary liquid, various ingredients were added to a 500 gallon mixer. The volume of the batch was 400 gallons which was heated to 40 °C to reduce the viscosity for stirring. After the mixing cycle had begun, five gallons of a proprietary additive was manually poured into the mixer through a two inch funnel/petcock arrangement, Figure 10.7. The additive was supplied in five gallon, rectangular carboys which were shipped in outer cardboard boxes. For reasons of stability the additive was kept refrigerated until used. The procedure was [1] obtain the carboy in its box from the refrigerator, [2] carry the box and carboy to the side of the mixer, [3] open the box and slide the carboy out of the box, [4] open the carboy by removing a screw cap, [5] cut a vent hole in the top of the carboy with a pocket knife, [6] open the petcock on the top of the mixer, and [7] pick up the carboy and pour its contents into the mixer.

In this incident ignition occurred just as the operator was beginning to pour the liquid from the carboy into the mixer. The explosion vented through the rupture disc into a knockout pot; but since the petcock was open at the time, a flame came into the workplace as well.

The investigation found that the outer cardboard box was made to fit tightly around the carboy. The inner surface of the box was therefore in intimate contact with the carboy during its movement around the workplace. This movement separated a equal and opposite charge on the surfaces of the box and the carboy. These charged surfaces neutralize each other as long as they remain in contact; but when the operator slid the carboy out of the box, there was charge separation between the box and the carboy. After removal, the outer surface of the carboy was highly charged. This surface charge induced a free charge onto the conductive liquid (cf., ¶ 2.11) which discharged to the funnel/petcock arrangement when the operator began to pour. There was a flammable vapor at the funnel which came from the flammable atmosphere within the mixer. The spark discharge occurred between the conductive liquid and the funnel to ignite the vapors coming from the mixer.

During the investigation another carboy was removed from the box and "looked at" with a field test meter. It was determined that the exterior of the carboy was more or less uniformly charged and there was a field strength of some 1×10^6 V/m at the surface; i.e., approximately one third of E_{max}; cf. ¶ 3.2.1. It was also determined that the relaxation time of the charge on the carboy was quite long, perhaps an hour or two if not more.

From Equation 11 a surface charge density is reckoned.

$$\sigma = \epsilon\epsilon_0 E = (1)(8.85 \times 10^{-12})(1 \times 10^6) = 8.85 \times 10^{-6} \text{ C/m}^2$$

The outer surface area of the carboy was some 0.4 m^2 so the total charge is easily determined.

$$(8.85 \times 10^{-6})(0.4) \approx 3 \times 10^{-6} \text{ C} = 3 \ \mu\text{C}$$

Thus a total charge of 3 μC was available for induction onto the conductive liquid in the carboy. Assuming the capacitance of the body of liquid to be some 10 pF (cf. ¶ 9.1.1), Equation 15 or Nomogram 9.2 yields a spark energy of some 450 mJ. The calculation of course assumes an ideal discharge from the liquid as if it were a perfect conductor, which it is not. Nevertheless, the calculation does show that there is orders of magnitude more energy in the system than that required to ignite flammable vapors.

As a result of the investigation the procedure of addition was modified.

10.8: Chemical Hose with an Ungrounded Spiral

An ignition of toluene vapors occurred in a glass lined reactor when the operator shut off the pump. The manway hatch to the reactor was closed but not dogged down. The combustion of the toluene vapors blew open the manway hatch and a flame issued forth into the workplace. There was minimal damage to the reactor and there were no personnel injuries.

Process Conditions:

Ambient conditions:
- 30° C (85 °F)
- 70% RH

Reactor:
- Volume: 3.78 m^3 (1,000 gallon)
- Internal temperature: ambient
- Pressure: ambient
- Contents: 3 m^3 (800 gallons) toluene

Toluene:
- Flash point: 4.4° C (40° F)
- Conductivity: undetermined (but cf., Table 6.1d)

Hose:
- Size: 1½" diameter in various lengths.
- Composition: rubber with three spiral stiffeners
- Couplings: metal
- (no contact between stiffeners and spirals!)

In a specialty chemical operation there were several mixers and reactors to which could be added a multitude of solvents and reagents.

The procedure was to connect a metering pump to one of several supply lines by means of a flexible hose. The output of the pump/meter was connected to a vessel by means of identical hoses and the transfer was started. When the proper amount of liquid had been metered into the vessel, as visually determined by the operator from the meter, the pump was turned off.

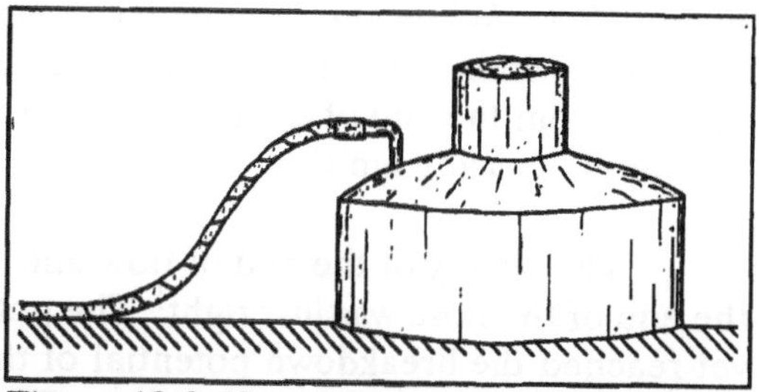

Figure 10.8: Hose Arrangement for Adding a Liquid to a Reactor

In the subject incident a batch was started by adding the proper amount of toluene to a glass lined reactor. The operator had completed the add and ignition occurred when he shut off the pump and was closing the valve at the output of the meter. Since he was well away from the reactor and there was no other personnel in the vicinity, there were no injuries. Only the undogged hatch cover was blown open and the open hatch vented the explosion. In this case the toluene vapors were fuel rich so that the maximum pressure and pressure rate were less than that of stoichiometric.

The flexible transfer hoses had been in use for some time and had been subjected to the rigors of the workplace. But more importantly, the specification for the hoses did not include that the spiral wire stiffeners be in positive metal to metal contact with the end couplings! An examination of the hoses showed that none of them showed electrical continuity between the couplings. Also many of them had become frayed where the hose mated to the coupling such that ends of the spirals were protruding through the rubber on the inside of the hose near the coupling; i.e., a spark gap.

As the toluene was pumped through the hose there was a streaming current - a charge flowing through the hose. This charge then accumulated on the ungrounded spiral wire. Since there is no electric field on the inside of the hose, the system behaves as a Van de Graaff generator (cf., ¶ 1.2) and charge would build up on the spiral wire until a breakdown potential was exceeded somewhere in the system; cf., ¶ 2.10.2 and Figure 2.7. In this case it is postulated that breakdown was occurring between an end of the spiral wire and a grounded coupling; the coupling being grounded by its metallic contact with the nozzle on the reactor,

Figure 10.8. While the toluene was being pumped the discharges were occurring within the liquid. But, when the pump was shut off toluene drained from the end of the hose at the reactor and discharge occurred in the vapors causing ignition.

The timing of the end of flow and the draining of the toluene from the end of the hose was just right. The potential on the spiral wire had not yet reached the breakdown potential of the toluene but had exceeded the breakdown potential of the air. Therefore, when the toluene/air atmosphere replaced the toluene liquid, discharge took place in the vapors.

As a result of this incident the specification for the hoses was changed to require that the spiral wires be attached to the couplings. However, there is an interesting sidelight to this specification since the flexible hoses have three separate spiral wires in them. The spirals are completely imbedded in the rubber of the hose so that they cannot be visually inspected to see that there is positive metal to metal contact at each end. The only check which can be made is making an electrical continuity check between the couplings. As the hoses are used to wearout; therefore, it is to be expected that sometime during the lifetime of the hose the spirals will become disconnected at the ends. When only one spiral becomes disconnected at both ends, an electrical continuity check cannot determine that there is an ungrounded spiral. Perhaps a better specification would be to have only one spiral in each hose and that the spiral be in positive metallic contact with the coupling.

10.9 Three Incidents in a Pneumatic Transport System

Process Description:

A granular product is produced in a spray drying process and sent to a surge hopper from which it is fed to a classifier through a screw feeder. The design intent of the classifier is to remove the fine material from the product to a specification of less than 3% through 200 mesh. The fine material from the classifier is recycled to the process via the classifier cyclone and the classifier bag filter. The coarse material from the classifier is pneumatically transported to the loading hopper in dense phase and subsequently dropped into an intermediate bulk container, IBC. The effluent pneumatic air from the process units go to their respective bag filters and on to atmosphere.

The exit chute of the loading hopper is equipped with vibrators to facilitate the movement of the powder into a wire reinforced flexible hose and on to the IBC. The flow of the material into the IBC is visually monitored and controlled by the operator by adjusting the opening of the

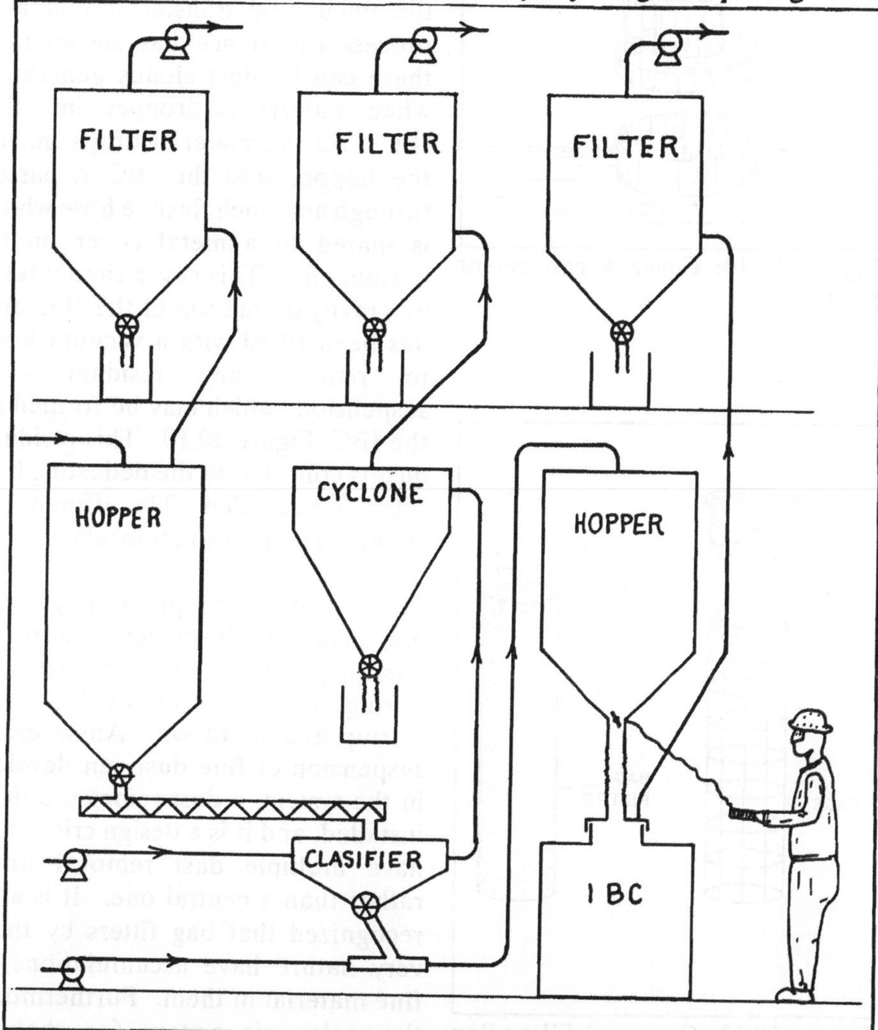

Figure 10.9: Example Pneumatic Transport System

butterfly valve in the throat of the hopper. When the IBC becomes full, the operator closes the butterfly and moves another IBC into position. A schematic of the process is shown in Figure 10.9.

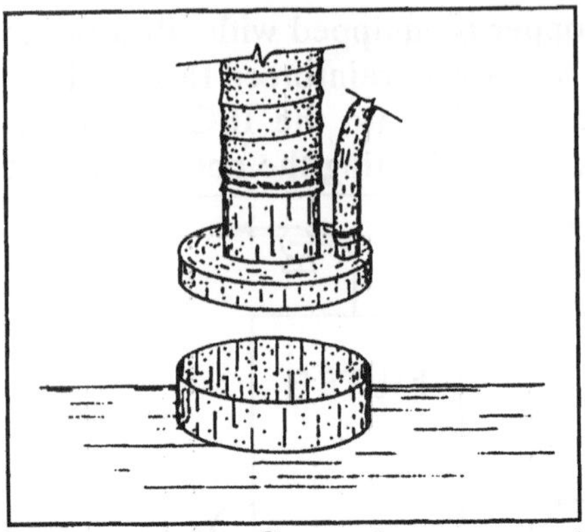

Figure 10.10: Cover Arrangement for IBC

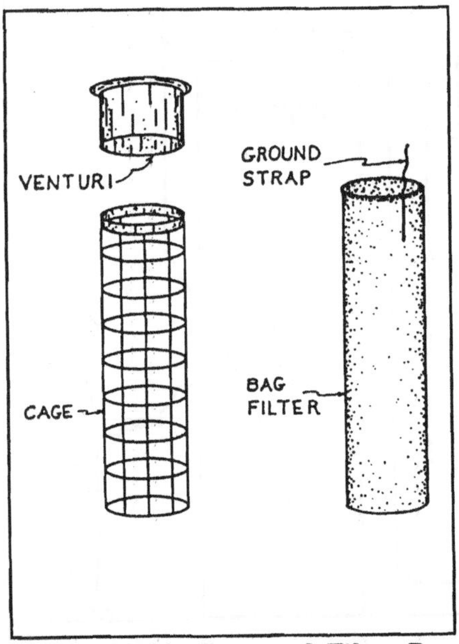

Figure 10.11: Cage and Filter Bag Arrangement

Since the classifier does not remove all of the fines and since there is attrition of some of the products as they are moved through the pneumatic transfer system and process experience has shown that there can be dust clouds generated when material is dropped into the IBC. As the material drops out of the hopper into the IBC it passes through an 8-inch flexible hose which is mated to a metal cover on its bottom end. This cover simply rests by gravity on the top of the IBC and has been fitted with a vacuum hose to remove any residual dust suspensions which may be formed in the IBC, Figure 10.10. This residual dust is collected in the dedusting bag filter and recycled. The effluent air stream is vented to atmosphere.

The design philosophy is to not have a fuel/oxidizer mixture (a dust suspension in air) in the process during normal operations, including startup and shutdown. Anywhere a suspension of fine dust can develop in the system, a dust removal unit is installed, and it is a design criteria to have multiple dust removal units rather than a central one. It is also recognized that bag filters by their very nature have accumulations of fine material in them. Furthermore, the pulse air system for shaking accumulated product from the bags necessarily creates a dust suspension when the pulses occur. It is therefore assumed that there will be dust suspensions present in the bag filters during normal operations. Since one cannot rely upon the absence of an ignition source to prevent explosions, the bag filters are equipped with explosion vents and located outside of the workplace.

To minimize parts inventory, all of the bag filter units are the same make and model. They are of the type where the bags are changed from the bottom of the tube sheet. The polyester bags have a copper ground strap woven into the lateral seam which extends from the seam so that it can be threaded over the top of the cage, out and under the band clamp, and attached to a grounding stud on the tube sheet; Figure 10.11. The design intent of the strap is to ground the cage, not the filter bag. The proper installation of the ground strap is essential in keeping all of the metal components if the unit properly grounded. When filter bags are used which have no ground straps, a separate ground wire is connected to the cage and threaded through the top to a grounding lug on the tube sheet.

Process Conditions:

System product throughput:	6,000 lbs/hr	2,722 kg/hr
Pneumatic flow rate:	200 scfm	5.66 m^3/min
Pneumatic pipe diameter:	3 in	76.2 mm
Volume of loading hopper:	450 cu ft	12.7 m^3
Mass in loading hopper, max.:	18,000 lbs	8,165 kg
Volume of IBC (4'x 4'x 6'):	96 cu ft	2.72 m^3
Mass in IBC, design intent:	3,600 lbs	1,633 kg
Ambient temperature:	80 °F	27 °C

Product Features: (measured)

Crystalline density:	1.58 gm/cc	1,580 kg/m^3
Nominal packing density:	40 lb/cu ft	640 kg/m^3
Hartmann, min. concentration:	0.03 oz/cu ft	0.03 kg/m^3
Hartmann, min. ignition energy:	10 mJ	10 mJ
Hartmann, max. pressure:	90 psi	6.2 bar
Hartmann, max. pressure rate:	9,500 psi/sec	655 bar/sec
Cloud ignition temperature:	842 °F	450 °C
Apparent dielectric constant:	2.1	2.1
Apparent conductivity:	10^{-12} S/m	10^{-12} S/m

Incident A:

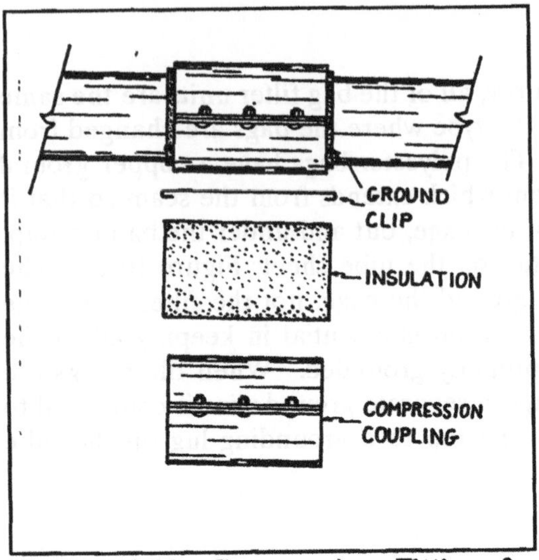

Figure 10.12: Compression Fitting for Pneumatic Transport Duct

During the initial shakedown of the system an operator heard a crackling noise when material was being transported from the classifier to the loading hopper. As he was investigating the source of the noise, he happened to touch a section of the duct and received a substantial electric shock. The system was immediately shut down and the duct was examined. It was found that a section of duct was not grounded because the grounding clips had not been installed at the unions of the ductwork section, Figure 10.12. The entire system was then checked for proper grounding and bonding of all metallic parts; and where deficiencies were found, corrections were made. This included all equipment units, all sections of ducts, and the bags and cages in the bag filters. The system was then turned back on and rechecked for any indication of electrostatic charge accumulation.

Incident B:

After eighteen months of operation an explosion occurred when an operator was filling an IBC with a standard product material. The loading hopper had been filled and product was being transferred from the hopper to an IBC through the butterfly valve and the flexible hose. The operator was regulating the flow of material by the hand held control unit which opened and closed the butterfly, Figure 10.9. The operator observed that the rate of flow began to decrease. This required an increase in the opening of the butterfly. When the operator had fully opened the butterfly control valve, the material began to flow rapidly into the IBC, such that the metal cover resting on the top of the IBC lifted and "danced about". At this time an explosion occurred and a fireball expanded into the workplace inflicting some minor burns on the operator.

The oxidizer for the explosion was, of course, the atmospheric oxygen in the hopper and the IBC. The fuel for the explosion was the product which had been dispersed as a dust within the IBC. In this case most of the fine material had been removed from the product stream by the classifier and by the loading hopper bag filter. Furthermore, the dedusting vacuum system on the IBC had removed some of the suspended dust. It was thought that all necessary precautions had been taken to eliminate the fuel (dust suspension) leg of the fire triangle from the IBC, but obviously such was not the case. There were additional circumstances, as discussed below, which had not been considered during the design phase HAZOPS analysis.

After the event, it could be seen that the explosion was not as violent as would have been predicted from the dust explosibility data. There was little damage to the equipment, the fireball did not propagate from the IBC to the dedusting bag filter, and there was no secondary fireball in the workplace. Apparently, the conditions for explosion were not optimum; the removal of the fines had been effective enough to reduce the amount of suspended fuel present so that there was inefficient combustion and a minimal explosion. However, the fireball was large enough to burn the operator who was standing near the IBC.

The accident investigation found that there had recently been some preventive maintenance performed on the system. The 8-inch diameter flexible hose between the loading hopper and the metal cover which rested on the IBC had become worn and was leaking. The 3-inch flexible hose between the metal cover and the metal pneumatic duct to the dedusting bag filter was also worn; new hoses were therefore installed. The 8-inch hose was made of a low conductivity rubber and had a spiral metal wire to give the hose rigidity. The spiral wire was completely imbedded in the rubber; however, when the millwright replaced the 8-inch hose, he simply cut a section of the required length and slipped it over the downspout from the loading hopper on one end and the collar of the metal cover on the other end. He made the hose fast with metal ring clamps, Figure 10.10. He made no effort to insure that the metal wire in the 8-inch flexible hose was bonded either to the loading hopper or to the metal cover; therefore, the wire in the 8-inch hose and the metal cover had no positive ground! The 3-inch hose was also made of a low conductivity rubber, but it had no metal spiral.

At the time of the incident all units were operating; the screw feeder was running, the classifier was running, and material was being fed to the loading hopper. An electrostatic charge was impressed on the material as it moved through the system and into the loading hopper. This electrostatic charge caused bridging in the loading hopper such that the material was not flowing smoothly into the IBC. As a result, the operator opened the butterfly in an attempt to increase the flow of the material. However, because of the bridging, the material ceased to flow even though the butterfly was fully opened. At this time there existed a dynamic situation of simultaneous electrostatic charging and dissipation. The charges were being created by the movement of the material into the loading hopper. At the same time, charges were being dissipated to the walls of the loading hopper: That is, the charges that created the bridge were being slowly dissipated so that the bridge became weaker with time. New material was being added to the top of the pile in the loading hopper and, even though it was charged, contributed to the sudden collapse of the bridge. When the bridge collapsed, a conglomerate of electrostatically charged material fell into the IBC. As the conglomerate was falling, the displacement of the air through the IBC caused the cover to "dance about". At this moment the metal cover was ungrounded and in the presence of an electric field from the falling, charged conglomerate. This caused an electrostatic charge to be induced onto the metal cover which then discharged in an incendive spark to the grounded IBC.

It is worthy to note that the bridging and the "dancing about" of the metal cover was an occurrence which had been noted before; however, it was considered to be an obvious way to vent the displaced air from the IBC.

Measurements made after the incident with an electronic insulation tester showed that the resistance between the metal cover and ground was somewhat greater than 10^{10} Ω; therefore, the metal cover was shown to be ungrounded if it was not touching the IBC. The IBC was metal and was sitting on a metal floor and a resistance measurement showed it to be adequately grounded.

It is estimated that the capacitance of the metal cover was of the order of 30 pF; therefore, the relaxation time, or RC time constant, for the cover is of the order of 0.3 seconds; i.e.,

$$t_r = RC \approx (10^{10})(30 \times 10^{-12}) \approx 0.3 \text{ second}$$

This estimation shows that the metal cover was capable of retaining the induced charge long enough to effectively produce the incendive discharge.

Some confidence in the scenario can be gained by some simple calculations using conservative assumptions. It is impossible to determine the amount of charge which was on the product at the time of the incident, but for products having such low conductivities (e.g., 1 pS/m), charge densities of 100 μC/kg can easily be attained during pneumatic transport (cf. Table 6.2). In this scenario the charge is moved into the loading hopper where it becomes compacted. As the charged powder becomes compacted there are mutual repulsive forces between the granules which exceed the forces of gravity. This results in the formation of a bridge of powder which impedes or stops the flow of powder from the hopper. But as the charge relaxes to the walls, the bridge will collapse when the forces of gravity take over. There is still a significant amount of charge remaining on the powder when it collapses and a conglomerate of powder falls into the IBC. The charge density on the conglomerate is unknown but 1 μC/kg is not unreasonable. As the conglomerate passes through the metal cover it is considered to be an equivalent cylinder 20 cm in diameter and 40 cm long of the process material having a density of 640 kg/m^3. Therefore, a charge of 8 μC being induced onto the metal cover can be reckoned; i.e.,

$$Q = \pi(0.1)^2(0.4)640 \times 10^{-6} \approx 8\ \mu C$$

This is an upper limit since a 100% efficiency for induction is assumed in the calculation, cf ¶2.11

Assuming the capacitance of the metal cover to be 30 pF and the charge induced upon it to be 8 μC, Equations 12 and 15 or Nomogram 9.2 can then be used to reckon a discharge energy and a potential.

$$W = Q^2/2C = (8 \times 10^{-6})^2/2(30 \times 10^{-12}) = 1.1\ J$$
$$V = Q/C = (8 \times 10^{-6})/(30 \times 10^{-12}) = 2.66\ kV$$

It is thus reckoned that a discharge energy of 1.1 joules and a potential of 266 kV would be possible based on the above conservative assumptions. In reality, discharge would occur before the potential of the metal cover reached 266 kV and the discharge energy would have been much less than 1.1 J, but it is shown that a discharge energy in excess of the experimental minimum ignition energy of 14 mJ is quite credible.

Incident C:

After two years of operation an explosion occurred in a bag filter. The bag filter was located on top of the roof with its explosion vents directed away from the building; therefore, there were no injuries to personnel when the fireball expanded into the atmosphere. The unit was rebagged, a new explosion vent was installed, and the unit was returned to service.

The investigation showed that the ground straps from the bags had not all been connected. Since some of the ground straps were still folded in with the bags, it was evident that the explosion itself did not disconnect the straps. Therefore, in the absence of other compelling evidence, an electrostatic scenario was postulated since all of the necessary components for electrostatic ignition were present. The discharge occurred between an ungrounded cage and some other grounded part of the unit, either an adjacent grounded bag and cage or the grounded frame of the unit. Discharge most probably occurred at the time of the pulse air cycle since this is when a dust suspension is created and the cages are physically moved about so they can touch one another or touch the grounded frame.

The creation of the spark separator in Incident A was simply that of introducing new technology into the workplace. Previously, the material had been handled manually and the need to keep everything grounded had not been a priority with the employees. When the new pneumatic equipment was installed the mechanics did not have the necessary awareness of proper grounding methods and neglected to always install the grounding clips, Figure 10.12. In this particular case, the incident did get the attention of the entire workforce to ground all metallic parts. It can therefore be said with some confidence that the initial state of the system was in accord with its design intent as far as electrostatics is concerned.

In incident B the product being manufactured was known to be susceptible to attrition in pneumatic transport. This prompted the installation of the dedusting bag filter. Even so, in this instance the probability of ignition of the dust cloud may have been low because the timing of all of the components of the conglomerate scenario had to be just right in order to obtain an incendive spark. Furthermore, all of the dust collection units were in place and working, but there was still the unrecognized conglomerate mechanism for obtaining a short lived dust suspension. At the time of the design review HAZOPS analysis, a dust suspension within the IBC was considered to be credible, and since it was the design philosophy not to have a dust suspension within the equipment, the dedusting bag filter was added to the system.

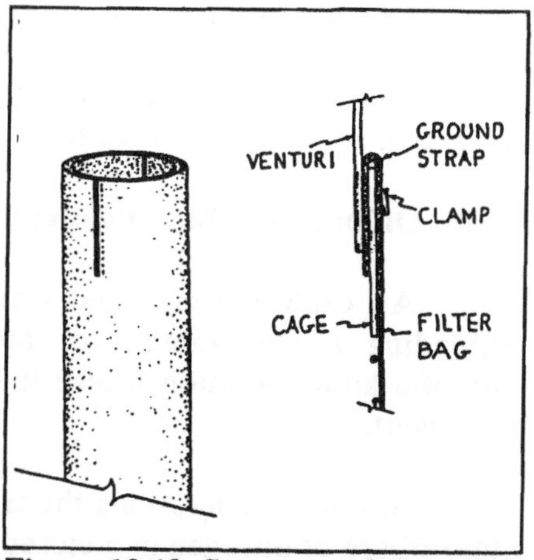

Figure 10.13: Suggested Grounding Strap Arrangement for Filter Bag

In Incident C, there was a situation in which a material with a low electrical conductivity was being moved through a pneumatic transport system where electrostatic charges were separated. These charges necessarily collected in the bag filter along with the product. It was the intent of the system to collect the fine portion of the material, the part which goes to make a dust explosion. The pneumatic transport fluid was air. Therefore, there was a fuel, an oxidizer, and some electrostatic energy which became the ignition source when it was collected on the ungrounded bags and cages. The bag filter had been rebagged several times during the course of the operations and the maintenance people had become somewhat complacent about keeping everything grounded.

Any time there is a fuel and an oxidizer in a process during normal operations, one cannot rely on the absence of an ignition source to prevent an explosion. Since it was known that a fuel (fine powder) and an oxidizer (air) would be present in the bag filters, they were placed on the roof and equipped with explosion vents which pointed away from the building. This design feature ameliorated the effects of the explosion in Incident C.

The ordinary construction of the ground straps on the filter bags is not conducive to the ease of maintaining a ground on the metal cage. The millwright must thread the ground strap through the cage and connect it to the stud on the tube sheet, an onerous task. Another way to maintain a ground on the cage has been suggested (D. Kirby, private communication). That is to sew two ground straps up the inside of the bag, over the top of the bag, and down the outside of the bag, Figure 10.13. By doing this, metal to metal contact between the venturi and the cage is assured. CAUTION: Some venturis are now made of plastic; this will prevent the establishment of a proper ground even when using an inside/outside grounding strap.

The lesson to be learned from these three incidents is that where electrostatic charges are ubiquitous in a process where flammable or combustible materials are also present, ungrounded conductors cannot be tolerated; sooner or later, things will come together to create an incident.

10.10: Offloading a Bulk Powder Truck

An explosion occurred when a bulk powder transport truck was offloading a consignment of atomized aluminum powder during a one-of-a-kind operation where the operators made up the procedure as they went.

In normal operations the truck had always been offloaded into an atmosphere of nitrogen in a closed hopper rail car. The exhaust from the diesel engine of the truck was the pneumatic transport fluid so that the rate of offloading created a concentration of aluminum powder which far exceeded the minimum concentration for a dust explosion. But since the exhaust was oxygen deplete the atmosphere in the hopper car had always been inert and there had been no problems.

In this instance, an order was canceled and the consignment of powder was sent back to the plant for offloading and reclassifying, an operation which had never been previously performed. The plan was to pneumatically move the powder from the truck to the entrance of the plant pneumatic transport system which was some distance away from where the truck could be parked. The three inch hoses on the truck could not reach the entrance so an additional hose of similar construction but of larger diameter was placed into service. The flexible hoses had metal fittings on each end and were made of rubber with a spiral of heavy wire within the rubber running between the flanges. In this manner an electrical connection was maintained with the truck, which in this case was properly grounded.

However, the flanges on the two different size hoses could not be connected. The operators devised a connection anyway by inserting the small hose into the larger one and stuffing rags between them to seal the opening, Figure 10.14. Thus, the last section of added hose was not grounded! The end of the larger, ungrounded hose was loosely placed into the entrance of the plant pneumatic transport system where the end could move about and bang against the wall of the grounded process equipment.

As if the ungrounded hose were not enough the operators added

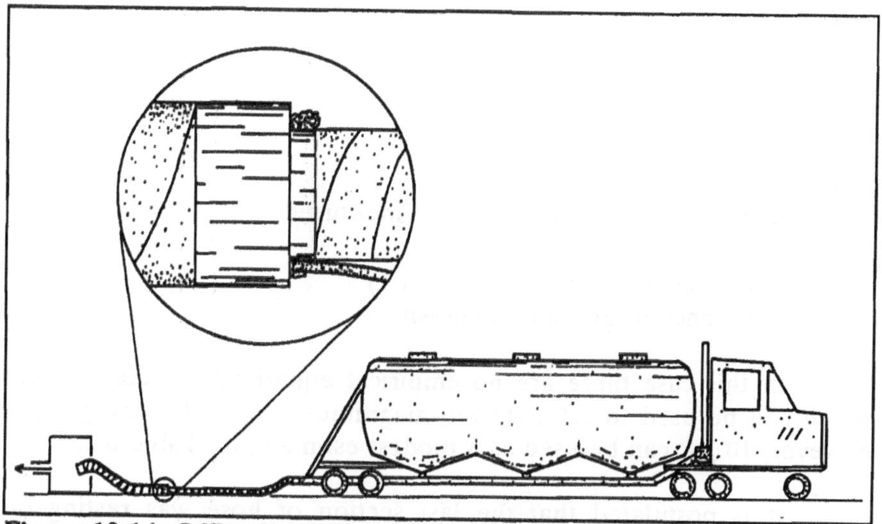

Figure 10.14: Offloading a Bulk Powder Truck

another element to the scenario. They recognized that the pneumatic transport system of the truck may not have been powerful enough to adequately move the product through the larger section of added hose. They therefore inserted a 3/8", high pressure, plant air hose into the opening between the couplings to "help things along". In so doing they defeated the inert characteristic of the pneumatic transport fluid.

In normal operation, the pneumatic transport system in the plant was operated in dilute phase well below the minimum concentration for ignition of the aluminum dust. The plant system was therefore operated with air as the transport fluid. Air was pulled into the plant system at its entrance where the truck hose had been inserted. The two pneumatic systems were therefore mismatched since the density of the aluminum powder being delivered by the transport medium of the truck was much greater than that of normal operations in the plant; i.e., an explosible dust-air mixture was inserted into the plant system.

The offloading operation was started and within a few minutes an aluminum dust explosion occurred which propagated throughout the plant. Because of the conditions, an electrostatic scenario for the ignition of the aluminum dust was considered.

Process conditions:

Aluminum:

Atomized, 10 micron, nominal
MIE: 50 mJ (Table 4.5)

Truck:

Resistance to ground: ~ 20 Ω (measured)

Additional hose:

Capacitance: ~ 30 pF (cf. ¶ 9.1.1)
Resistance to ground: unknown

In this case there are no empirical equations or rules of thumb which can be used to estimate the magnitude of the streaming current; however, 10 μA can be used as a modest estimate, cf. Table 6.2.

It is postulated that the last section of hose was resting on a concrete floor with only a rubber/concrete interface; therefore, the resistance to ground from the spiral wire should be somewhat in line with the values for resistivity given in Table 9.1. Furthermore, similar hoses have shown resistances to ground of $> 10^{10}$ Ω when tested with an insulation tester. Thus, a reasonable estimate of the resistance to ground of the last section of hose is $10^{10} > R > 10^{9}$ Ω.

The steady state voltage which could have been impressed on the section of ungrounded hose can then be derived from Ohm's Law, Equation 1.

$$V = I_sR = 10^{-5} \times 10^{10} = 100{,}000 \text{ V}$$

$$V = I_sR = 10^{-5} \times 10^{9} = 10{,}000 \text{ V}$$

Then, with a capacitance of 30 pF a discharge energy can be reckoned from Equation 13 or Nomogram 9.2.

$$W = \tfrac{1}{2}CV^2 = \tfrac{1}{2}(30 \times 10^{-12})(10^{5})^2 = 0.15 \text{ J} = 150 \text{ mJ}$$

$$W = \tfrac{1}{2}CV^2 = \tfrac{1}{2}(30 \times 10^{-12})(10^{4})^2 = 0.0015 \text{ J} = 1.5 \text{ mJ}$$

These results compare with the MIE for atomized aluminum at some 50 mJ. Thus, a modest estimate for the streaming current, capacitance of the hose, and its resistance to ground shows that an electrostatic discharge is a very credible scenario for the incident.

10.11 Dumping Powder from a Drum with a Metal Chime

An incident occurred when an operator was dumping a "dusty" powder from a polyethylene drum. The drum had been equipped with a metal chime to withstand the rigors of the workplace. The fixed receiver into which he was dumping the powder was made of metal and was well grounded through the process equipment, Figure 10.15.

As the material was dumped it slid down the interior of the plastic drum and into the receiver. As it fell it also created a dust cloud. During dumping, the metal chime on the lip of the drum was ungrounded and free standing. After dumping the operator withdrew the drum toward himself and touched the metal chime to the fixed metal receiver. As this metal gap was closed, an incendive spark occurred which ignited the dust cloud.

Figure 10.15: Dumping a Powder from a Polyethylene Drum with a Metal Chime

In the manipulation of powders, frictional charging occurs at the interfaces with process equipment and at intergranular interfaces. A quantitative measure of the amount of charge which can be separated on a powder can be obtained by simulating the process environment and dumping some powder into an instrumented Faraday cage. This amount of charge can then accumulate on pieces of ungrounded process equipment from which discharges can occur.

Process conditions:

Powder:

- Amount dumped: 18.1 kg (40 lbs.)
- Conductivity: 2.5 10^{-12} S/m (measured)
- Dielectric constant: 4.1 (measured)
- Minimum ignition energy: 12 mJ (measured)

Drum:

Body of drum: Polyethylene, PE
Diameter: 0.406 m (16 in.)
Height: 0.508 m (20 in.)
Minor diameter of chime: 0.016 m (0.63 in.)
Capacitance of chime (metal): 71 pF
Dielectric constant of PE: 2.3 (Table 9.1)
Bulk conductivity of PE: 1 10^{-9} S/m (measured)
Surface resistivity of PE: 1 10^{12} Ω (measured)

A process drum and 40 pounds of process material were obtained. The material was dumped into an instrumented Faraday cage where it was determined that a total charge of 3.6 μC was separated on the powder during the dumping operation. An equal and opposite amount of charge was assumed to remain on the plastic drum. An attempt was made to check this assumption by the use of a field test meter. (The drum was too large to insert into the available Faraday cage.)

The charge remaining on the drum was bound to the surface of the polyethylene; and since the surface had such a low conductivity, the charge was not evenly distributed. The field test meter indicated almost zero charge on the surface opposite where the powder slid into the Faraday cage. In the area where the sliding took place, the field test meter indicated its maximum value of 500 kV/m. In any case the charge bound to the surface of the polyethylene was induced onto the metal chime. The surface area of the chime was derived by assuming the geometry of a torus.

$$A = 4\pi^2 r_1 r_2 = 4\pi^2(0.008)(0.203) = 0.0641 \text{ m}^2$$

If the total charge, 3.6 x 10^{-6} C, could have been induced on this surface, 0.0641 m^2, it would have exceeded the maximum charge density which can be held on a surface in air; i.e., 27 $\mu C/m^2$, cf., ¶ 3.2.1;

$$\sigma = Q/A = 3.6 \times 10^{-6} / 0.0641 = 5.6 \times 10^{-5} \text{ C/m}^2.$$

If one assumes a high efficiency for inducing a charge on the chime, cf., ¶ 2.11, the charge on the chime will be limited by corona discharge since $\sigma > \sigma_{max}$ to the maximum value of 27 $\mu C/m^2$. Notice that the simple dumping of some process material has created the maximum charge which can be held on the surface of a piece of ungrounded process equipment!

The energy in the discharge can be derived from the surface of the chime, the maximum charge density, and the capacitance of the chime.

$$Q = \sigma_{max}A = 27 \times 10^{-6} \times 0.0641 = 1.7 \times 10^{-6}\ C.$$

$$W = Q^2/2C = (1.7 \times 10^{-6})^2/2(71 \times 10^{-12}) = 0.020\ J = 20\ mJ$$

This compares with the minimum ignition energy for the dust at 12 mJ which shows that the ignition of the dust by electrostatics is a credible scenario. Charge was separated by frictional charging in the dumping of the powder, the charge was accumulated by induction of the bound charge onto the ungrounded chime, and discharge occurred when the chime approached the grounded process equipment.

10.12 Emptying a Powder from a Plastic Bag. (Composite Case History)

The Manufacturing Chemists Association published "Case Histories of Accidents in the Chemical Industry" during the 1960's. At that time MCA members anonymously submitted a brief writeup of incidents as a means of sharing safety information. Unfortunately, these were discontinued when the MCA changed its charter and its name to the Chemical Manufacturer's Association. The CMA has even thrown out their archive copies of the histories! Nevertheless, there are extant copies of the case histories which are quite useful in obtaining actual examples of incidents.

Noteworthy among them are incidents #203, #627, #700, #958, #969, #1094, #1180, & #1817 which all have in common the emptying of a powder from a plastic bag into a flammable atmosphere where ignition occurred. The following is a composite of these incidences.

An employee was emptying a 25 pound fiber drum of an organic resin into a reactor containing methyl alcohol. He had removed the lid from the drum and opened the plastic bag used as an innerliner. He carefully folded the bag back over the rim of the drum and dumped the majority of the resin into the reactor by picking up the drum and tipping it against the lip of the reactor. After he placed the drum back down on the floor, he removed the plastic innerliner and was in the process of shaking the last of the resin into the reactor when ignition of the alcohol vapors occurred. (Since this case is a composite, the extent of damage is not specified but operators have been severely burned in similar instances.)

Process Conditions:

Resin:

Amount dumped: 11.3 kg (25 lbs., typical)
Conductivity: 3 x 10^{-12} S/m (typical)
Dielectric constant: 4 (typical)
Minimum ignition energy: 10 mJ (typical)

Plastic bag:

Volume: 0.014 m^3 (typical)
Conductivity: 1 x 10^{-14} to 1 x 10^{-18} S/m (Table 9.1)
Dielectric constant: 2.3 (Table 9.1)

Methyl alcohol (methanol):

Minimum ignition energy: 0.14 mJ (Table 4.1)
Volume in reactor: 1.0 m^3 (typical)

Reactor:

Volume: 1.5 m^3 (typical)
Temperature: 25°C (77°F)

Dean, et. al. (1992) developed a satisfying mathematical model for the man/bag equivalent circuit in the dumping operation. The analysis of their model showed that if the charge accumulated on the ungrounded man, there was no question that there was enough energy to ignite ordinary vapors. On the other hand if the man were grounded, then their model showed that the discharge could have come from the bag. They validated their analysis with an experimental program where they demonstrated that a surface charge density of 1.8 x 10^{-5} C/m^2 is to be expected from the dumping of a typical powder from a generic plastic bag. (Several plastics were tested.) This exceeds the surface charge density requirement for incendive brush discharge of the order of 10^{-5} C/m^2 found by Gibson and Harper (1981).

Gibson and Harper (1988) showed that flammable atmospheres with MIEs of 0.2 mj and 0.04 mj can be ignited respectively by discharges from 10 cm^2 and 4 cm^2 areas of electrostatically charged polyethylene sheets, and that plastic pipes of radius greater than 0.5 cm can ignite 0.2 mj atmospheres. In their experiments, Dean et. al. (1992) showed that the charged surface area of the bags used were approximately 0.159 m^2. (It should be appreciated that the entire 0.159 m^2 area is not discharged into the spark, but that there is much more area than that required.)

The experiments showed that humidity played a significant role in whether enough charge was separated onto the plastic substrate. In this context it must be appreciated that it is the humidity within the bag that is the controlling factor in an industrial situation, not the humidity of the workplace. If the material was packed and kept dry, as many materials are, then the humidification of the operating area may be a questionable fix.

The process conditions reported in the MCI incidents (and others) are quite variable and the exact scenarios are different, but charge separation occurs at two places. At the interface between the innerliner and the drum as the full drum is moved about prior to being emptied and between the resin and the plastic bag as the resin slides from the drum into the reactor. These two can be additive. Charge is accumulated on the inside of the bag when the resin separates and carries with it the opposite charge. Charge is accumulated on the outside of the bag when the operator removes the bag from the drum leaving the opposite charge on the inside of the drum.

In the composite example the minimum ignition energy for methyl alcohol is much lower than that of the resin so one would expect that it was the alcohol vapors which ignited. There is however the incident #1094 with charcoal dust where, apparently, no vapors were involved.

10.13 Vapor Explosion in a Closed Tank

This Case History was presented as a paper by the author at the AIChE 27th Annual Loss Prevention Symposium and was subsequently published in *Process Safety Progress*, 12 (1993) 203-205.

An explosion occurred inside a 1,000 gallon tank which was being used to hold treated wastewater. A light fraction of kerosene carried over from the previous step was attributed to be the fuel for the combustion reaction. The wastewater contained hydrogen peroxide which was decomposing and led to an oxygen enriched atmosphere in the head space. The tank had been standing in a quiescent state for some two hours before the explosion occurred. An additional observation was made by an operator shortly after the explosion occurred. The operator stated that "the liquid was bubbling like seltzer". This bubbling was most probably oxygen bubbling to the surface as a result of the decomposition of hydrogen peroxide. This observation can lead one to speculate on a novel electrostatic scenario.

Under many process conditions an aqueous phase can contain impurities such that the vapor space above the liquid will be flammable. If the liquid phase also contains a peroxide, the peroxide can decompose to form gaseous oxygen. When this occurs, bubbles of oxygen form and rise to the surface where they burst. When the bubbles break upon arriving at the surface, small droplets of the liquid are ejected to form a mist. Therefore, under favorable conditions, a flammable atmosphere of a fuel (mist and vapor) in 100% oxygen is formed at the surface of the liquid. The entire head space of the tank will not be at 100% oxygen but the atmosphere right at the surface will be.

The liquid droplets dispersed into the atmosphere upon the bursting of the bubbles also carry a significant electrostatic charge, such that a space charge will be formed in the head space of the tank. As charging continues, the gradient at the surface increases until corona discharge occurs at a small bit of conductive matter floating on the surface which acts as a collector. This corona discharge then ignites the flammable atmosphere at the surface and the flame propagates throughout the head space.

Corona discharge is not incendive to ordinary flammable vapors in air. (Hydrogen, acetylene, and carbon disulfide are among the exceptions.) Data for ignition of vapors by corona discharge in oxygen atmospheres is sparse at best, (This translates as the author has not found any), but there are some data for MIEs of a few materials in both air and oxygen atmospheres at normal conditions of pressure and temperature. These data show that the energy requirement for ignition by capacitive spark discharge is some 100 times less in an oxygen atmosphere than in air, Table 4.4. It is assumed then that the requirements for ignition by corona discharge are likewise much less, and that ignition of ordinary flammable vapors by corona discharge in oxygen would be expected.

The scenario of a corona discharge ignition at the surface of the liquid is consistent with the two hour rest time in the cited incident since the liquid was observed to be bubbling after the explosion occurred.

However, if this mechanism is indeed viable, then the question arises as to why such ignitions have not been observed many times before. Perhaps the answer lies in the notion that so many things must come together at the right place and the right time that the probability of their simultaneous occurrence is small. Since the details of the process are ill defined, it is not possible to enumerate those parameters whose effects would be expected to be significant. Nevertheless, surface tension, viscosity, conductivity, dielectric constant, double layer thickness, polarity, flash point, and minimum ignition energy would be among the ones to consider.

10.14 Gas Well and Pipeline Blowouts

There have been a number of incidents where there has been a blowout of a two phase stream of flammable gas/liquid and ignition of the resulting vapor cloud. A definitive reference (Young, 1960) is given for an ignition at a gas well blowout after water had been injected into an offset well. There have been other cases where the ignition of vapor clouds from blowouts were attributed to electrostatics: [1] A gas well blowout off the coast of Louisiana on September 29, 1992 (USCG, 1994), [2] A gas well blowout under a workover rig in Midland County Texas on March 8, 1993, and [3] An underwater blowout in Lake Maracaibo, Venezuela under Flow Station EF37 on January 2, 1994.

The postulated mechanism is that the high velocity escaping of a gas/liquid stream from a pipe is a horrendous separator of electrostatic charge. The resulting gas cloud is both flammable and highly charged. In all of the above cases there was metallic, grounded equipment in the gas cloud which constituted a site for brush discharge. Therefore, the four conditions for electrostatic ignition are present: separation, accumulation, discharge, and a flammable atmosphere.

In these cases it is postulated that the rate of charge separation by aerosol formation is greater than the rate of charge dissipation by the expansion of the gas cloud and that the buildup of the electric field is great enough to cause brush discharge at a grounded piece of process equipment. The competing forces of mutual electrostatic repulsion between the aerosol droplets and the mobility of the aerosol droplets through the air are such that the expansion of the cloud is sufficiently retarded to allow the buildup of an electric field of sufficient magnitude to cause brush discharge.

Getting a quantitative handle on this phenomena would be a rather extensive theoretical and experimental undertaking. In the meantime, the postulated mechanism will have to suffice for the explanation of the known occurrences..

Appendix A

Units

The International System of Units (SI) has been adopted for the text. This system is the modern version of the MKSA (meter, kilogram, second, ampere) system and ASTM E 380 has been used as the basic practice for the usage of the units and their abbreviations.

Meter: The meter is the SI base unit of distance and has the symbol **m**.

Kilogram: The kilogram is the SI base unit for mass and has the symbol **kg**.

Second: The second is the SI base unit for time and has the symbol **s**.

Ampere: The ampere is the base electrical unit in the SI system from which all other electrical units are defined through the use of definitions, the basic laws of physics, and derived mathematical relationships. The ampere is the unit of current flow, and for a constant current it is one coulomb per second. The ampere is defined in the International System of Units, SI, as "*The ampere is that constant current which, if maintained in two straight parallel conductors of infinite length, of negligible cross section, and placed one meter apart in vacuum, would produce between these conductors a force equal to* 2×10^{-7} *newton per meter of length.*" The SI symbol for the ampere is **A**.

Volt: The volt is the SI unit of potential and is defined in terms of potential energy per charge, or joules per coulomb. (A free standing conductor holding one coulomb of charge at a potential energy of one joule will be at a potential of one volt.) The SI symbol for the volt is **V**.

Ohm: The ohm is the SI unit of resistance to current flow and is defined in terms of amperes per volt. (It takes a potential of one volt to push a current of one ampere through a resistance of one ohm.) The SI symbol for the ohm is **Ω**.

Siemens: The siemens is the SI unit of conductance and is defined as a reciprocal ohm. (Conductance is the reciprocal of resistance.) The SI symbol for the siemens is **S.**

Coulomb: The coulomb is the SI unit for a quantity of charge. One electron carries an elementary charge, e, of 1.602×10^{-19} coulombs; therefore, the charge carried by 6.24×10^{18} unit charges is one coulomb. (A current of one ampere flowing for one second delivers one coulomb of charge.) The SI symbol for the coulomb is **C.**

Farad: The farad is the SI unit of capacitance and is defined as charge per potential or coulombs per volt. (A one farad capacitor holding one coulomb of charge will be at a potential of one volt.) The SI symbol for the farad is **F.**

Joule: The joule is the SI unit of energy and is defined in mechanics as a newton·meter. Coulomb's Law has been defined so that in electrical units it is a volt·coulomb. (It takes one joule of energy to push one coulomb of charge across a potential of one volt.) The SI symbol for the joule is **J.**

Newton: The newton is the SI unit of force. A force of one newton on a mass of one kilogram will result in an acceleration of one meter per second per second. The SI symbol for the newton is **N.**

Pascal: The pascal is the SI unit of pressure. A force of one newton per square meter is one pascal. The Si symbol for the pascal is **Pa.**

A	Ampere	N	Newton
C	Coulomb	Pa	Pascal
F	Farad	S	Siemens
J	Joule	s	second
kg	kilogram	V	Volt
m	meter	Ω	Ohm

Appendix B

Symbols Used in Equations

- A Area, m^2
- C Capacitance, F
- E Field strength, V/m
  - E_n Normal field strength
  - E_b Dielectric breakdown strength
- I Current, A
  - I_s Streaming current
  - I_{ch} Charging current
  - I_d Discharge current
- P Pressure, Pa
- Q Charge, C
- R Resistance, Ω
- S Volume charge density, C/m^3
- T Temperature, °C
- V Potential (Voltage), V
  - V_b Breakdown voltage
- W Energy, J

- a Linear dimension, m
- b Linear dimension, m
- d Diameter, m
- h Height, m

- k Empirical coefficient, $m^3/C{\cdot}s$ (Equation 24)
- ℓ Length, m
- m Coefficient, $°C^{-1}$ (Equation 23)
- n Ionic mobility, $m^2/V{\cdot}s$ (in Bustin equation, Equation 22)
- r Radius, m
- t Time, s

- ϵ Dielectric constant, dimensionless
- ϵ_0 Permittivity of a vacuum 8.85×10^{-12} F/m
- ζ Zeta potential, V
- η Viscosity, Pa·s
- κ Conductivity, S/m
- λ Surface conductivity, Ω
- ρ Resistivity, Ω·m
- σ Surface charge density, C/m^2
- τ Relaxation time, s
- υ Velocity, m/s

Appendix C

Equations

Mathematical relationships are used in the text as the occasion warrants. They are repeated in the running text to save the student the onerous task of looking them up each time they are used. They are referred to by the numbers given here in the hopes that it will save the student a trip to his old physics book.

Ohm's Law: Ohm's states that the amperage through a conductor is directly proportional to the electric potential across that conductor.

$$I = \frac{V}{R} \tag{1}$$

I ≡ Current, A
V ≡ Potential, V
R ≡ Resistance, R

Resistivity: Resistivity is an intrinsic property of a material and is defined in terms of the resistance, length, and cross section of a body of the subject material. The resistance of a conductor of constant cross section is directly proportional to its length and inversely proportional to its cross sectional area. The proportionality constant is the resistivity of the material from which the conductor was made.

$$R = \rho\frac{\ell}{A} \tag{2}$$

ρ ≡ Resistivity of material, Ω·m
R ≡ Resistance of a conductor made from the material, Ω
ℓ ≡ Length of the conductor, m
A ≡ Cross sectional area of the conductor, m^2

Conductivity: Conductivity is the reciprocal of resistivity

$$\kappa = \frac{1}{\rho} \tag{3}$$

κ ≡ Conductivity, S/m

Surface Resistivity: Surface resistivity is an intrinsic property of a material's surface, if pure, and is defined in terms of the resistance to the flow of current across a rectangle of the surface in question. Surface impurities can alter its value significantly.

$$R = \lambda \frac{\ell_1}{\ell_2} \tag{4}$$

λ ≡ Surface resistivity of a material, Ω (or ohms per square)
R ≡ Resistance across a surface of the material, Ω
ℓ_1 ≡ Length of the surface, m
ℓ_2 ≡ Width of the surface, m

Potential: There is an energy associated with bringing a test charge into an electric field. The energy required to bring a test charge into an electric field is a force-distance integral and the force which is exerted at any given point will be in direct proportion to the charge on the test charge. The potential at any point in an electric field is then defined as the energy per unit charge required to bring the test charge to that point. The potential at any point from a point charge is directly proportional to the charge and inversely proportional to the radius from the point charge.

$$V = \frac{Q}{4\pi\epsilon\epsilon_0 r} \tag{5}$$

V ≡ Potential, V
Q ≡ Charge, C
$\epsilon\epsilon_0$ ≡ Permittivity, F/m
r ≡ Radius from point charge, m

The symbol for potential is V and since it is defined as energy per charge it has the units of joules per coulomb; however, by the laws of physics and the corresponding transposition of units it is expressed in units of volts. Thus, "potential" and "voltage" are synonymous.

Field Strength: An electric field is a region in space where electric forces can be experienced. Field strength is a measure of the intensity of an electric field at a point in space and is defined in terms of the force per unit charge exerted on a very small test charge placed in the electric field; i.e., if a charged body is present in an electric field there will be an electric force upon it, and the more intense the field - the greater the force. The field strength at any point from a point charge is directly proportional to the charge and inversely proportional to the square of the radius from the point charge.

$$E = \frac{Q}{4\pi \epsilon \epsilon_0 \, r^2} \tag{6}$$

E ≡ Field strength, E
Q ≡ Charge, C
$\epsilon\epsilon_0$ ≡ Permittivity, F/m
r ≡ Radius from point charge, m

The symbol for field strength is **E** and since field strength is defined in terms of force per unit charge it has the units of newtons per coulomb; however, by the laws of physics and the corresponding transposition of units it is expressed in units of volts per meter, V/m, and is the gradient of voltage in an electric field.

$$E = -\frac{\partial V}{\partial \ell} \tag{7}$$

E ≡ Field strength, V/m
V ≡ Potential, V
ℓ ≡ Length, m

Note that **E** is a vector quantity having both magnitude and direction.

Permittivity: Coulomb's law states that there will be an electrostatic force between two charged spheres which is directly proportional to the charge on each sphere and inversely proportional to the square of the distance between their centers. The proportionality constant used when all quantities are in SI units includes the permittivity of the medium in which the spheres are suspended. The baseline medium is that of a vacuum and the symbol for the permittivity of a vacuum is ϵ_0 where

$$\epsilon_0 = 8.85 \times 10^{-12} \text{ F/m}$$

When ϵ_0 appears in Culomb's law the units are coulombs squared per newton meter squared but in other relationships it has the equivalent units of farads per meter, seconds per ohm meter, coulombs per volt meter, etc

When conductors are separated by a medium other than a vacuum and the medium will pass an electric field, the medium is termed a dielectric; i.e., a dielectric is a substance that will pass an electric field. The permittivity of a dielectric is denoted by $\epsilon\epsilon_0$ where ϵ is the dielectric constant; i.e. the dielectric constant is a ratio of the permittivity between the dielectric and a vacuum. Note that its magnitude is always greater than one and that the dielectric constant of a good conductor, eg. metal, is infinite.

Relaxation Time: If a charged capacitor has a path of some resistance through which the charge can dissipate, the relaxation of the charge will be a first order exponential decay.

$$Q = Q_0 e^{\left[-\frac{t}{\tau}\right]} \tag{8}$$

Q ≡ Charge on capacitor at time t, C
Q_0 ≡ Charge on capacitor at zero time, C
t ≡ time, s
τ ≡ Relaxation time, s (Time to reach 1/e of initial charge)

The relaxation time can also be expressed in terms of the capacitance and the resistance of the circuit.

$$\tau = RC \tag{9}$$

R ≡ Resistance of the circuit, Ω
C ≡ Capacitance of the circuit, F

The relaxation time for a charge to dissipate from a material can be expressed in terms of the material's resistivity or its conductivity.

$$\tau = \rho \epsilon \epsilon_0 = \frac{\epsilon \epsilon_0}{\kappa} \tag{10}$$

ρ ≡ Resistivity of the material, Ω·m
κ ≡ Conductivity of the material, S/m
$\epsilon\epsilon_0$ ≡ Permittivity of the material. s/Ω·m

Surface Charge Density: The strength of an electric field at the surface of a charged surface is directly proportional to the surface charge density.

$$\sigma = \epsilon \epsilon_0 E \tag{11}$$

σ ≡ Surface charge density, C/m^2
$\epsilon\epsilon_0$ ≡ Permittivity, C/V·m
E ≡ Field strength, V/m

Capacitance: When two conductors in the same vicinity are given equal amounts of charge of opposite polarity, the arrangement is termed a capacitor, i.e., an arrangement where charge can be stored. The capacitance of a capacitor is the ratio of the charge on either conductor to the potential difference, voltage, between them.

$$C = \frac{Q}{V} \tag{12}$$

C ≡ Capacitance, F
Q ≡ Charge, C
V ≡ Potential, V

Energy: The energy stored in a capacitor is given by the following relations:

$$W = \frac{1}{2}CV^2 \tag{13}$$

$$W = \frac{1}{2}QV \tag{14}$$

$$W = \frac{Q^2}{2C} \tag{15}$$

W ≡ Energy, J
C ≡ Capacitance, F
Q ≡ Charge, C
V ≡ Potential, V

Streaming Current: In a material handling operation when a charged substance flows (streams) through a pipe there is necessarily a current streaming through the pipe. An empirical relationship has been developed (Schön, 1965) for the streaming current.

$$I_s = 3.75 x 10^{-6} \upsilon^2 d^2 \tag{16}$$

$$S = 4.77 x 10^{-6} \upsilon \tag{17}$$

More recently, Britton (1988) suggested more conservative relationships.

$$I_s = 2.5 x 10^{-5} \upsilon^2 d^2 \tag{18}$$

$$S = 3.18 x 10^{-5} \upsilon \tag{19}$$

I_s ≡ Streaming current in liquid, A
S ≡ Volume charge density of liquid in pipe, C/m^3
υ ≡ Velocity of liquid in pipe, m/s
d ≡ Pipe diameter, m

It is not of much practical value, but a correction term for short pipes has been suggested for the empirical streaming current relations.

$$\left(1-e^{\left[-\frac{\ell}{\upsilon\tau}\right]}\right) \tag{20}$$

ℓ ≡ Length of pipe, m
υ ≡ Velocity of liquid in pipe, m/s
τ ≡ Relaxation time of liquid, s

Years ago Helmholtz (1879) gave a classic derivation of the streaming current in a liquid flowing through a pipe.

$$I_s = \frac{A\epsilon\epsilon_0\zeta\Delta P}{\eta\ell} \tag{21}$$

a ≡ Cross section of pipe, m^2
$\epsilon\epsilon_0$ ≡ permittivity of liquid, F/m
ζ ≡ Zeta potential, V
ΔP ≡ Pressure drop, Pa
η ≡ Viscosity of liquid, Pa·s
ℓ ≡ Length of pipe, m

Hyperbolic Decay: For liquids having conductivities below one picosiemen per meter, charge decay should be reckoned from the Bustin equation. (Bustin, 1983) rather than from the exponential relationships of Equations 8-10.

$$S = \frac{S_0}{\left[1 + \frac{nS_0t}{\epsilon\epsilon_0}\right]} \tag{22}$$

S_0 ≡ Initial charge density, C/m_0
n ≡ Ionic mobility, $m^2/V{\cdot}s$
t ≡ Time, s
$\epsilon\epsilon_0$ ≡ Permittivity of the liquid, $s/\Omega{\cdot}m$

Conductivity with Temperature: The conductivity of an insulative liquid decreases with a decrease in temperature.

$$\log_{10}\left(\frac{\kappa_1}{\kappa_2}\right) = m(T_1 - T_2) \qquad (23)$$

κ_1 ≡ Liquid conductivity at temperature T_1, pS/m at °C
κ_2 ≡ Liquid conductivity at temperature T_2, pS/m at °C
m ≡ Log conductivity/temperature coefficient, $°C^{-1}$

The range of 0.009 - 0.0018 for the coefficient m has been suggested for aviation fuels (Gardner and Moon, 1983)

Charge Decay from Mists: A mist can hold a space charge and the rate at which the space charge decays is governed by coalescence. Bustin (1983) has shown that the decay follows the following relationship:

$$S = \frac{S_0}{1 + S_0 kt} \qquad (24)$$

S ≡ Charge density at time t, C/m^3
S_0 ≡ Initial charge density, C/m^3
t ≡ Time from beginning, s
k ≡ Empirical constant, $m^3/C{\cdot}s$ ($2 - 6 \times 10^4$ $m^3/C{\cdot}s$)

Capacitance: The capacitance of a free standing sphere is directly proportional to its radius.

$$C = 4\pi\epsilon\epsilon_0 r \qquad (25)$$

r ≡ Radius of sphere, m
$\epsilon\epsilon_0$ ≡ Permittivity of the medium surrounding the sphere, F/m
C ≡ Capacitance, F

The capacitance of a parallel plate capacitor, neglecting fringing effects, is directly proportional to the area of the plates and inversely proportional to the spacing between them.

$$C = \frac{\epsilon\epsilon_0 A}{\ell} \tag{26}$$

$A \equiv$ Area of plates, m^2
$\ell \equiv$ Distance between the plates, m
$\epsilon\epsilon_0 \equiv$ Permittivity of dielectric material between the plates, F/m
$C \equiv$ Capacitance, F

Appendix D

Atmospheric Electrostatics

Nutshells:

[1] One typical ordinary lightning bolt releases approximately 3×10^{8} joules of energy. One lightning superbolt releases around 5×10^{11} joules (Turman, 1979).

[2] The peak optical power from ordinary lightning bolts approximate a log normal distribution with median power of 10^{9} watts and standard deviation of 10.8 decibels
(Turman, 1978). The energy conversion efficiency for optical radiation is about 0.4% (Krider et al., 1968). Cianos and Pierce (1972) also report a log normal distribution with the median current in a first return stroke of lightning at 20 kiloamperes, 2×10^{4} A.

[3] Current flowing across the surface of the earth into a lightning strike can easily create voltage drops of 1,000 volts per meter 100 meters from the point of the strike (Wahlin 1986).

[4] Cosmic radiation creates approximately 10^{7} ion pairs per cubic meter per second at the earths surface. The average lifetime of a small ion is about 100 seconds; therefore, the average population of small ions in air is about 10^{9} ion pairs per cubic meter (Wahlin, 1986).

[6] The average fairweather current density at the surface of the earth is about 4×10^{-12} amperes per square meter (Wahlin, 1986).

[7] The average fairweather field strength at the surface of the earth is about 100 volts per meter and decreases to less than 10 volts per meter at an altitude of 3 kilometers (Wahlin, 1986).

10.1 Lightning

There is no doubt that there is enough energy in a lightning strike to cause ignition of any of the fuel/oxidizer mixtures capable of undergoing combustion. Questions do however arise about the conditions required to get ignitions by indirect means; e.g., "The lightning strikes over there but will ignition occur over here?" or "Can ignitions occur during periods of atmospheric electrical activity?"

There is an upper limit to which the forces of nature can charge a thundercloud. When this limit is approached, the awesome phenomena of lightning occurs. Cloud to ground lightning usually occurs from a negatively charged, low altitude cloud to the literal ground of the earth. However, about one strike in 100 comes from a positively charged cloud and there are the rarer super bolts which come from positive, high altitude clouds. (Williams 1988)

In the usual lightning from negative clouds, there is a level of charge density in the lower portion of the cloud at which a discharge called a step leader exits from the bottom of the cloud. This step leader forms a channel of ionized air and it takes a bit of time for the resistance of the channel to lower sufficiently for the ionization to continue. This is not a continuous buildup of an ionized channel but a process which proceeds in a stepwise fashion; thus the term "step leader". The formation of the channel cannot usually be seen with the naked eye as it steps down toward the earth in steps of approximately 50 meters. Meanwhile, the temporal and spatial configuration of the accompanying electric field is changing as the charged channel descends downward. When the channel gets to within 50 to 100 meters from the earth the electric field is so intense between the lower end of the channel and the earth that extremely large gradients are created. At this point positive discharges called streamers are drawn off of all sharp points in the area under the end of the step leader. There may be many streamers created in this process and all of them will be moving to connect up with the end of the step leader. The higher and the sharper the point, the sooner the streamer will connect up with the step leader. Only one of the streamers will win the race and it will then begin to drain off charge from the lower part of the step leader. This reduction of charge in the step leader causes the electric field at the surface of the earth to be significantly reduced and the other streamers collapse. When the step leader and the streamer meet, a continuous channel of ionized air is formed between the charged cloud and the earth. This channel is conductive and provides the path through which charge, or current, can flow. This is the main channel of the first return lightning stroke where current rushes from the ground to the cloud creating a brilliant flash of light and thunder.

All of the charge in the cloud is not neutralized by this first return stroke and there may be several subsequent return strokes of a lower magnitude. About 50% of the first return strokes carry a current of 20 kiloamperes, about 1% exceed 200 kiloamperes, and auout 99% exceed 2 kiloamperes (Cianos and Pierce 1972).

When the first return stroke occurs, tremendous amounts of current are drawn from the vicinity of the strike. There may we!l be configurations of process equipment in unprotected units where localized arcing can occur and cause ignitions; it is therefore prudent to provide proper lightning protection in those units where flammables, combustibles, or explosives may be present. See NFPA 78.

10.2 Fairweather Current

Atmospheric electrical activity is not confined to periods of thunderstorms. Cosmic radiation from the far reaches of the universe is continuous and unabating. When these cosmic particles collide with the molecules in the air, electrons are knocked out of their orbits and molecular ions are formed. Those molecules which have an electron removed are positive ions and those molecules which attach themselves to the free electrons are negative ions. Positive and negative ions have different mobilities (migration rates) and can therefore become separated. This separation is one part of the mechanism that makes the atmosphere surrounding the earth act as a huge battery. This results in an electric field around the earth which is known as the fairweather field. Associated with it is a fairweather current between the negative surface of the earth and the excess positive ions in the atmosphere. These currents and fields are of such low magnitude that they do not lead to problems of ignition of flammable mixtures or explosives.

10.3 Ignitions by Fairweather Electrostatics?

There is however an area of gray between the extremes of thunderstorm activity where there can certainly be problems and fairweather activity where no ignition problems are known to exist. It is known that electric fields at the surface of the earth will increase at the approach of incipient storms, which may or may not develop into thunderstorms, and that similar increases remain as storms and thunderstorms abate, sometimes for extended periods; however, experience in this area shows that further concern about inadvertent ignitions is unwarranted.

Appendix E

Electric Field Calculations

It is sometimes instructive to calculate electric field parameters within a tank containing a charged liquid. These sorts of calculations get rather onerous in a hurry, but a case where the mathematics have been worked out is that of assessing an electric field within a partially full tank. Carruthers and Wigley (1962) have published formulas for a rectangular tank (ship's tanks are rectangular) and Bustin (1983) has extracted two useful relations from their work.

In assessing an electrostatic situation within a tank one may need to relate the charge in the liquid to the potential at the liquid surface. Alternatively the surface potential may have to be calculated from a field strength measured at the roof with a fieldmeter. As always there are some assumptions which must be made. In this case it is that the charge is uniformly distributed within the liquid and that there are no internal structures within the tank which will distort the electric field. With these assumptions in place the potential at the free liquid surface can be reckoned from Equation E2 and the field strength (voltage gradient) at the center of the tank roof can be reckoned from Equation E3

$$V = \sum_r \sum_s \frac{576\, Q\, (-1)^{\frac{r+s+2}{2}}}{\pi r s \beta^2} \frac{[\cosh(\beta d) - 1]\, \sinh(\beta C - \beta d)}{\sinh(\beta c) + (\epsilon - 1)\cosh(\beta d)\sinh(\beta c - \beta d)} \qquad \text{(E1)}$$

$$E = \sum_r \sum_s \frac{576\, Q\, (-1)^{\frac{r+s+2}{2}}}{\pi r s \beta} \frac{[\cosh(\beta d) - 1]}{\sinh(\beta c) + (\epsilon - 1)\cosh(\beta d)\sinh(\beta c - \beta d)} \qquad \text{(E2)}$$

r and s are summed over odd integers 1, 3, 5, ...
Q ≡ charge density in liquid (nC/m^3)
ϵ ≡ dielectric constant of liquid
a ≡ tank width, m
b ≡ tank length, m
c ≡ tank height, m
d ≡ liquid height, m

$$\beta = \pi\sqrt{\left(\frac{r}{a}\right)^2 + \left(\frac{s}{b}\right)^2}$$

The relationships above can provide guidance in some practical situations. When more sophisticated calculations are required, one may want to go to the references. In the case of cylindrical geometry, eg. round tanks and drums, the derivations of Britton (1988) are suggested.

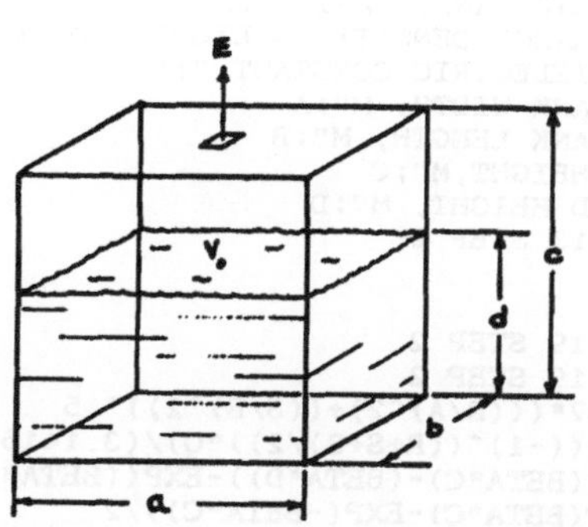

Figure E1: Parameters for a Rectangular Tank Partially Full of a Charged Liquid.

GW Basic Program for Equation E1

```
10 REM BUSTIN'S ELECTROSTATIC FIELD CALCULATION
20 REM CALCULATION FOR POTENTIAL AT CENTER OF LIQUID
21 REM RECTANGULAR TANK, PART FULL
30 INPUT "Q = CHARGE DENSITY IN LIQUID, nC/cuM ";Q
40 INPUT "E = DIELECTRIC CONSTANT ";E
50 INPUT "A = TANK WIDTH, M";A
60 INPUT "B = TANK LENGTH, M";B
70 INPUT "TANK HEIGHT,M";C
80 INPUT "LIQUID HEIGHT, M";D
100 FOR Z=1 TO 13 STEP 2
120 NEXT Z
150 W=0
160 FOR R=1 TO 19 STEP 2
190 FOR S=1 TO 19 STEP 2
200 BETA=3.14157*(((R/A)^2)+((S/B)^2))^.5
210 GAMMA=(576*((-1)^((R+S+2)/2))*Q)/(3.14157*R*S*(BETA^2))
220 SINHF=(EXP((BETA*C)-(BETA*D))-EXP((BETA*D)-(BETA*C)))/2
230 SINHBC=(EXP(BETA*C)-EXP(-BETA*C))/2
240 COSHBD=(EXP(BETA*D)+EXP(-BETA*D))/2
250 DELTA=((COSHBD-1)*SINHF)/(SINHBC+((E-1)*COSHBD*SINHF))
255 IF S>13 GOTO 270
270 W=W+(GAMMA*DELTA)
280 NEXT S
300 NEXT R
310 PRINT
320 PRINT  " POTENTIAL = ";INT(W);"VOLTS AT CENTER OF THE LIQUID SURFA
CE"
330 END
```

GW Basic Program for Equation E2

```
10 REM BUSTIN'S ELECTROSTATIC FIELD CALCULATION
20 REM CALCULATION FOR GRADIENT AT CENTER OF ROOF
21 REM RECTANGULAR TANK, PART FULL
30 INPUT "Q = CHARGE DENSITY IN LIQUID, nC/cuM ";Q
40 INPUT "E = DIELECTRIC CONSTANT ";E
50 INPUT "A = TANK WIDTH, M";A
60 INPUT "B = TANK LENGTH, M";B
70 INPUT "TANK HEIGHT,M";C
80 INPUT "LIQUID HEIGHT, M";D
100 FOR Z=1 TO 13 STEP 2
120 NEXT Z
150 W=0
160 FOR R=1 TO 19 STEP 2
190 FOR S=1 TO 19 STEP 2
200 BETA=3.14157*(((R/A)^2)+((S/B)^2))^.5
210 GAMMA=(576*((-1)^((R+S+2)/2))*Q)/(3.14157*R*S*(BETA))
220 SINHF=(EXP((BETA*C)-(BETA*D))-EXP((BETA*D)-(BETA*C)))/2
230 SINHBC=(EXP(BETA*C)-EXP(-BETA*C))/2
240 COSHBD=(EXP(BETA*D)+EXP(-BETA*D))/2
250 DELTA=(COSHBD-1)/(SINHBC+((E-1)*COSHBD*SINHF))
255 IF S>13 GOTO 270
270 W=W+(GAMMA*DELTA)
280 NEXT S
300 NEXT R
310 PRINT
320 PRINT  " GRADIENT = "; :PRINT USING "###.##";W/1000;
325 PRINT "  KILOVOLTS/METER AT CENTER OF ROOF"
330 END
```

Bibliography

American National Standard for Personal Protection - Protective Footwear, ANSI Z41-1991, American National Standards Institute, Inc., 1430 Broadway, New York, NY 10018

American Petroleum Institute, "Protection Against Ignitions Arising Out of Static, Lightning, and Stray Currents" *API Recommended Practice RP2003*, Fifth Edition, March 1991.

Bailey A. G., "Electrostatic Hazards in Powder Silos", *Inst. Phys. Conf. Ser. No.85*, 1987, 1-13.

Bartknecht W., *DUST EXPLOSIONS*, Springer-Verlag, New York, ISBN 0-387-50100-2, 1989.

Berkey B. D., T. H. Pratt, and G. M. Williams, "Review of Literature Related to Human Spark Scenarios", *Plant/Operations Progress, Vol. 7, No. 1*, Jan. 1988, 32-36.

Blanchard D. C., "The Electrification of the Atmosphere by Particles from Bubbles in the Sea", *Progress in Oceanography*, Vol.1; M. Sears, Editor; The Macmillan Company, New York, 1963.

Blythe A. R. and W. Reddish, "Charges on Powders and Bulking Effects", Inst. Phys. Conf. Ser. No. 48, *Electrostatics 1979*, 1979, 107-114.

Bordutha F. T., *Industrial Explosion Prevention and Protection*, McGraw-Hill Book Company, New York, 1980. ISBN 0-07-006359-1.

Boschung B., W. Hilgner, G. Lüttgens, B. Mauer, and A. Widmer, "An Experimental Contribution of the Question of the Existence of Lightning-Like Discharges in Dust Clouds", *J. of Electrostatics, 3*, 1977, 303-310.

British Standards Institution *Code Of Practice for Control of Undesirable Static Electricity*, BS 5958, Part 1. General Considerations, 1991; and Part 2. Recommendations for Particular Industrial Situations, 1991.

Britton L. G., "Using Material Data in Static Hazard Assessment", *Plant/Operations Progress, Vol. 11, No. 2*, April 1992, 56-70.

Britton L. G. and J. A. Smith, "Static Hazards of Drum Filling", *Plant/Operations Progress, Vol. 7, No.1.* Jan. 1988, 53-78.

Britton L. G., "Systems for Electrostatic Evaluation in Industrial Silos", *Plant/Operations Progress, Vol. 7, No. 1*, Jan. 1988, 40-50.

Britton L. G., "Static hazards using flexible intermediate bulk containers for powder handling", *Process Safety Progress*, 12 (1993) 240-250.

Britton L. G. and T. J. Williams, "Some characteristics of liquid-to-metal discharges involving a charged 'Low Risk' oil", *J. Electrostatics*, 13 (1982) 185-207.

Brundrett G.W., "A Review of the Factors Influencing Electrostatic Shocks in Offices" *J. Electrostatics, 2*, 1976/1977, 295-315.

Bustin W. M., I. Koszman and I. T. Tobye, "New Theory for Static Relaxation from High Resistivity Fuel", American Petroleum Institute, Division of Refining, Vol. 44 [III] (1964)

Bustin W. M. and W. G. Dukek, *ELECTROSTATIC HAZARDS IN THE PETROLEUM INDUSTRY,* Research Studies Press Ltd., Lecthworth, England, 1983. ISBN 0 86380 000 9.

Calcote H. F., C. A. Gregory, Jr., C. M. Barnett, and Ruth B. Gilmer, "Spark Ignition: Effect of Molecular Structure", *Ind. and Eng. Chem. 44, No. 11,* 1952, 2656-2662

Carruthers, J. A. and K. J. Wigley, "The Estimation of Electrostatic Potentials, Fields, and Energies in a Rectangular Metal Tank Containing Charged Fuel", *J. Inst. Petrol.48* (1962) 180.

Cross J. A., *ELECTROSTATICS: PRINCIPLES, PROBLEMS AND APPLICATIONS,* Adam Hilger, Bristol, England, 1987, ISBN 0-85274-589-3.

Dean J.C., D.A. Hulburt, A.l. Matthews and G. M. Williams, "Quantitative Evaluation of Electrostatic Hazards", Paper from International Conference on Hazard Identification and Risk Analysis, Human Factors and Human Reliability in Process Safety, held in Orlando, FL, January 15-17, 1992 by HSE, AIChemE and Eur. Fed. Chem. Eng., pp.85-105.

Eckoff R. K., *DUST EXPLOSIONS*, Butterworth-Heinemann 1991, ISBN 0 7506 2007 2.

Eichel F. G., "Electrostatics", *Chemical Engineering*, Mar. 13, 1967. McGraw-Hill Inc.

Electrostatic Discharge Association, *Electrostatic Discharge Control Handbook*, Electrostatic Discharge Association, Rome, NY 13440, 1994

ESCIS, Expert Commission for Safety in the Swiss Chemical Industry, "Static Electricity, Rules for Plant Safety", *Plant/Operations Progress, Vol. 7, No. 1*, Jan. 1988, 1-22.

Fedoroff B. T. and O. E. Sheffield, Editors, *Encyclopedia of Explosives and Related Items,* Picatinny Arsenal, Dover, New Jersey, 1972

Finke J. "Electrostatic Effects of Charged Steam Jets", *J. Electrostatics*, 23 (1989) 69-78.

Gardner L. and F. G. Moon, "The relationship between electrical conductivity and temperature of aviation turbine fuels containing static dissipator additives", National Research Council Canada, NRC No. 22648, 1983.

Gavis J. and J. P. Wagner, "Electric Charge Generation During Flow of Hydrocarbons through Microporous Media." *Chemical Eng. Sci., 23* 381-391, 1968

Gavrilenko N. D., "Minimum ignition energies for vapor-gas-air mixtures at various temperatures", Unpublished.

Gibson N., "Electrostatic Hazards with Liquids and Powders" Lecture: University of Alexandria Research Center, Alexandria, Egypt, 1981.

Gibson N. and F. C. Lloyd, "Incendivity of Discharges from Electrostatically Charged Plastics", *Brit. J. Appl. Phys. 16*, 1965, 1619-1631.

Gibson N. and D. J. Harper, "Evaluation of Electrostatic Hazards Associated with Non-Conducting Materials", *J. Electrostatics 11*, 1981, 27-41.

Gibson N. and F. C. Lloyd, "Static Electricity in Chemical Plant and

Processes" *Technical Memorandum, RD6653A*, Imperial Chemical Industries, Ltd., Dyestuffs Division, May 27, 1963.

Glor M. "Discharges and hazards associated with the handling of powders", *Inst. Phys. Conf. Ser.*, No. 85, Section 3, Electrostatics '87, 207-216.

Glor M. "Ignition of gas/air mixtures by discharges between electrostatically charged plastic surfaces and metallic components" *J. Electrostatics*, 10 (1981) 327-332.

Haase H., *ELECTROSTATIC HAZARDS: THEIR EVALUATION AND CONTROL*, Verlag Chemie, Weinheim, New York, 1977.

Heidelberg E., "Generation of igniting brush discharges by charged layers on earthed conductors", *1967 Static Electrification Conf.* 147-155.

International Safety Guide for Oil Tankers and Terminals, Third Edition, International Chamber of Shipping, Oil Companies Marine Forum, International Association of Ports and Harbors, Printed in England by Witherby & Co. Ltd., London, EC1, 1988.

Johnson J. K., "The Ignition of Vapour and Droplets by Liquid-to-Metal Sparks", *J. Electrostatics, 4*, (1977/1978) 53-65.

Jones T.B. and S. Chan, "Charge Relaxation in Partially Filled Vessels", *J. Electrostatics*, 22 (1989) 185-197.

Katrak K. K., "Human body electrostatic charge (ESC) levels: Are they limited by corona bleed off or environmental conditions?", EOS/ESD Symposium 1995, Paper 2.3, 1995.

Klinkenberg A. and J. L. van der Minne, *ELECTROSTATICS IN THE PETROLEUM INDUSTRY: THE PREVENTION OF EXPLOSION HAZARDS*. A Royal Dutch/Shell Research and Development Report, Elsevier Publishing Co., 1958.

Kramer H. and K. Asano, "Incendivity of sparks from surfaces of electrostatically charged liquids", *J. Electrostatics, 6* (1979) 361-371.

Leonard J. T., and H. F. Bogardus, "Pro-static Agents in Jet Fuels", Naval Research Laboratory Report 8021, August 16, 1976, (NTIS AD A029382)

Lewis B., and G. von Elbe, *COMBUSTION, FLAMES AND EXPLOSIONS OF GASSES*, Second Edition, Academic Press, New York, 1961.

Lüttgens G., "Collection of Accidents Caused by Static Electricity", *J. Electrostatics 16*, 1985, 247-255.

Lüttgens G. and M. Glor, *UNDERSTANDING AND CONTROLLING STATIC ELECTRICITY*, Expert Verlag GmbH, Goethestrasse 5, 7044 Ehningen bei Boblingen, Federal Republic of Germany, 1989. ISBN 3-8169-0510-2.

Mancini R. A., "The Use (and Misuse) of Bonding for Control of Ignition Hazards", *Plant/Operations Progress Vol. 7 No. 1,* January 1988, 23-31.

Manufacturing Chemists Association, Inc., "Case Histories of Accidents in the Chemical Industry."

National Fire Protection Association, *NFPA 77 Recommended Practice on Static Electricity,* 1993 Edition, National Fire Protection Association, Quincy, MA 02269.

National Fire Protection Association, *FIRE PROTECTION HANDBOOK*, Sixteenth Edition, 1986, National Fire Protection Association, Quincy, MA 02269.

Nifuku M., B. Vonnegut, and D. C. Blanchard, "Charged Drops Produced by Bursting of Bubbles at the Surface of Organic Liquids", *J. Electrostatics*, 2 (1976) 279-282.

Perry J. H., Editor, *CHEMICAL ENGINEERS' HANDBOOK*, Third Edition, McGraw-Hill, New York, 1950.

Perry R. H., and C. H. Chilton, Editors, *CHEMICAL ENGINEER'S HANDBOOK*, Fifth Edition, McGraw-Hill International Book Company, ISBN 0-07-049478-9, 1974

Schön G. and H. Kramer, "On the size of stationary space charge clouds in streaming media", *Static Electrification, 1971*, Conference Series Number 11, Paper 14, 138-152, Proceedings of the third conference on static electrification organized by the Static electrification Group of The Institute of Physics held in London, May 1971.

Thomas G. O., G. Oakley, and J. Brenton, "Influence of the Morphology of Lycopodium Dust on Its Minimum Ignition Energy", *Combustion and Flame 85*, 526-528 (1991).

Tucker T. J., "Spark Initiation Requirements of a Secondary Explosive", pp643-653, *Annals New York Academy of Sciences*, Volume 152, Art. 1, "Prevention of and Protection Against Accidental Explosions of Munitions, Fuels and Other Hazardous Materials." October 28, 1968.

Turman B. N., "Lightning Detection from Space" Source unknown.

United States Coast Guard Marine Casualty Investigation Report, Case Number MC92016479, 03AUG94, Casualty of the vessel Blake IV on 29SEP92.

von Pidoll U., H. Kramer, and H. Bothe, "Avoidance of ignition of gasoline/air mixture during refueling of motor vehicles at filling stations." DGMK Deutsche Wissenschaftliche Gesellschaft für Erdöl, Erdgas und Kohle E. V., DGMK Project 508, ISSN 0937-9762, 48 pages, 16 figures, 26 references, 1996

Wahlin L., *ATMOSPHERIC ELECTROSTATICS,* Research Studies Press Ltd., Lecthworth, England, 1986.

Walmsley H.L., "The Avoidance of Electrostatic Hazards in the Petroleum Industry", *J. of Electrostatics, 27,* Nos 1 & 2, January 1992. ISSN: 0304-3886.

Williams E. R., "The Electrification of Thunderstorms" *Scientific American*, November 1988 88-99.

Young J. W., R. A. Hemstock, J. T. Middleveen, and D. P. Howell, "Electrostatics in Gas Blowouts", *The Canadian Mining and Metallurgical Bulletin 53, No.581*, 1960, 682-686.

CONCORDANCE A

GENERAL

CONCORDANCE B

COMPOUNDS and MATERIALS

Printed and bound by CPI Group (UK) Ltd, Croydon, CR0 4YY

Printed and bound by CPI Group (UK) Ltd, Croydon, CR0 4YY

16/09/2026

14963420-0001